Ma

636.9   Turner, Isabel
3234   Exhibition and pet cavies / by Isabel
Tur   Turner. -- 2nd ed. -- Hindhead, Surrey :
        Saiga, 1981
            vii, 152 p. : ill.

        1. Guinea pigs 2. Guinea pig breeds I. Title.

SEP 2 0 1983

EXHIBITION AND PET CAVIES

BOOKS AVAILABLE IN THE SERIES

Exhibition and Pet Rabbits
Meg Brown

Exhibition and Pet Mice
Tony Cooke, LRIC

Exhibition and Practical Goatkeeping
Joan Shields

Exhibition and Flying Pigeons
Harry G. Wheeler

Book of the Netherland Dwarf
Denise Cumpsty

Exhibition and Pet Hamsters and Gerbils
David Robinson

Exhibition and Pet Cavies

by

ISABEL TURNER S.R.N. R.F.N.

SECOND EDITION

PUBLISHED BY:
SAIGA PUBLISHING CO. LTD.,
1 Royal Parade, Hindhead, Surrey,
England GU26 6TD.

Printed in Great Britain
by Photobooks (Bristol) Ltd
and bound by
Western Book Company Ltd.

Published by
SIAGA PUBLISHING CO. LTD.,
1 Royal Parade, Hindhead, Surrey,
England GU26 6TD.

Contents

Acknowledgements

My grateful thanks to all my fancier friends for their encouragement, and to those who so willingly supplied me with photographs, with special mention of Prof. H. Elvidge, of the Nuffield Medical Research Institute of Oxford, and my good friend Cathy Whiteway who keeps me right with Genetics, and Zena Goodman for her help with typing and her helpful criticism.

I should also like to thank Janice Lawson-Reay and Rachel Gibbons for providing photographs for the second Edition.

List of Illustrations

Introduction

THE ORIGIN OF THE DOMESTIC CAVY

When the Spanish Conquistadors landed in South America nearly 500 years ago, they found that the Incas had domesticated the wild cavies, and had, either accidentally or deliberately through selective breeding, produced colours other than the wild agouti colour. The main colours they had bred were white, and tricolours in black, red and white—the forerunners of our tortoise-shell and whites of to-day.

The Incas allowed these small animals to run around their homes and fed them on scraps from their own table. But although they were treated as pets, they were sacrificed and eaten on special occasions. In order to prepare them for cooking the hair was scraped off, as the cavy is very tough-skinned and difficult to skin, and when they were prepared for cooking they looked like tiny pigs and the Spaniards called them "Cochinillo das Indas" or the "little Indian pig". The Incas called them "cui". It is not known how the "guinea" was added, but one theory is that the Spaniards transported them to Guiana, and from there the Dutch merchants took them to Europe as pets for their children.

In more modern times in South America, the Indians still keep them in their homes, and still eat them on special occasions, but they have lost the knack or the interest in breeding different colours, and most of their cavies are dirty whites and dingy browns. The Indians offer their cavies for sale in their markets but they are not a very economic proposition. In more recent years, the Department of Agriculture in Peru have done a great deal of research, in an effort to breed bigger and better cavies, and so make them an economical source of meat.

THE GUINEA PIG AS A PET
AND FOR EXHIBITION

So, when they arrived in Europe, they became very popular as children's pets, as evidenced by the fact that there must be many millions of pet guinea-pigs being kept at any time. Then, as mutations occur, which produce different colours, coat type and other characteristics different varieties evolve, and so a Fancy is born, each fancier trying to breed "better" cavies.

This involves setting up specialist breed clubs as well as a National, and regional clubs. *Standards*; i.e. acceptable descriptions for judges and breeders to follow, have to be established. There have to be facilities for showing and exchanging information. All this and much more come together to make a fascinating hobby for young and old alike.

CHAPTER 1

Housing

There are many different types of hutches, some better than others, but if you want your cavies to thrive they must be housed adequately.

Ideally, if you intend to keep cavies in numbers, you should have a shed specially for them—a caviary. This is more comfortable for the cavies and yourself when attending to them. If you are a good handyman, you can make your own hutches to fit your shed.

SIZE OF HUTCHES

Floor space should be at least one and a half square feet for one cavy; twice that size would house a pair, or a sow with a litter. Large runs are useful for several litters of weaned youngsters, or resting sows. All should be at least eighteen inches high.

Wood is the best material for hutches, but it absorbs moisture, so it is a good idea to fit a piece of wood in the form of a tray to cover the base of the hutch. This piece can then be renewed when necessary. Wire floors are *not* satisfactory for cavies. Make your hutches so that the complete front opens; this will make it easy to remove the tray and clean the hutch. A piece of wood about two

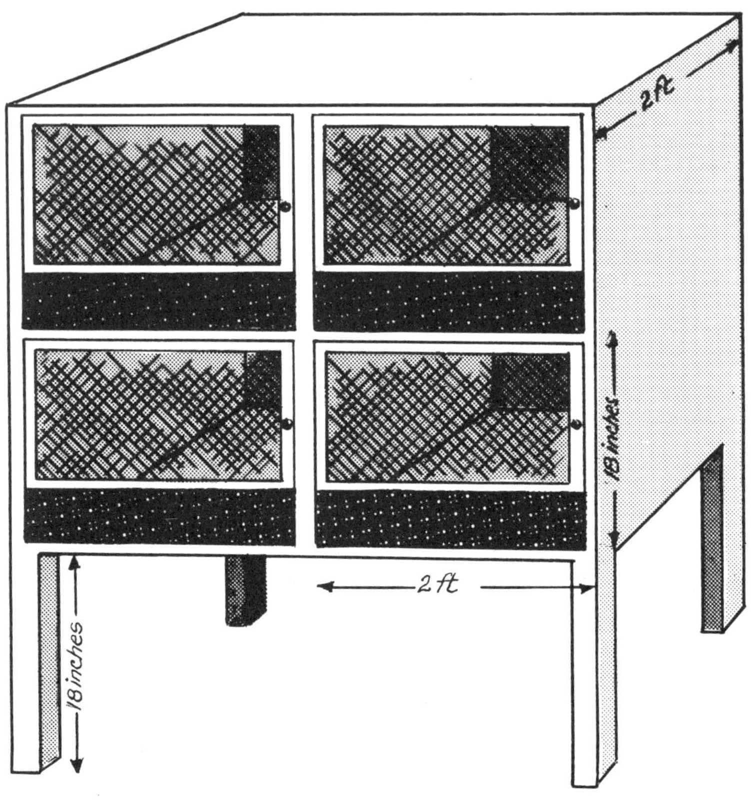

Block of four indoor hutches to house single cavies.

Dark shaded panels are boards across the front of the hutch to keep the cavy from falling out when the door is open.

inches high, to fit across the open front of the hutch, but removable for cleaning will ensure that your cavies do not fall out when you open the hutch door. If your hutches are in a shed, then the front may be entirely wire netting, but if they have a large frontage it is better if the front is part wood.

Do not have your bottom hutches at floor level, or you will find it back-breaking to feed and clean them. About a foot above floor-level is ideal. Also, for comfort, do not have the top ones too high. Hutches too close to the ceiling of the shed can become very hot in summer, and could kill your cavies.

Some fanciers have only a slat of wood across the opening in the hutch, and no netting wire, but I have always felt that the cavies are safer when they are closed in. With an open front, there is always the risk of a rat or a stray cat getting in, and though most cavies will not climb over the barrier, sometimes a bolder one will fall out. Also for safety make a wire netting door to fit over the door opening of the shed, when you want to leave the door open during hot weather. A similar piece of wire netting on a frame may be fixed over the window when it is open.

No artificial heating is necessary, but the shed should be weather-proof and draught-proof. Small animals require fresh air, but must be protected, especially when there is frost, snow or rain.

OUTDOOR HUTCHES

If you cannot manage to get a shed, or if you need some extra hutches, it may be necessary to have outdoor

continued on page 8

2ft
6ft
3ft

MORANT HUTCH

A large Morant hutch which can house a colony of cavies. The closed-in part should have a wooden floor. The open part should have a wire netting floor, so that the cavies can graze. The sides or ends can be hinged to open to allow access, and the Morant can be lifted and moved to another site as required.

hutches. They should be more substantial than the indoor ones—able to stand up to bad weather, and to give adequate protection from the weather to your cavies. They should have a separate sheltered compartment, with an opening into the main part of the hutch to give the cavies easy access. This gives them somewhere to seek warmth from the cold, and shade from the sun in the hot weather. The roof, sides and backs of the hutches should be covered with roofing felt so that no damp can get in and a shutter should be fixed over the wire netting section at night. The shutter should leave about an inch clear space at the top for ventilation.

To make things more comfortable for yourself, rig up some sort of overhanging shelter over your outdoor hutches, so that you have some shelter from inclement weather, when tending your cavies.

So far, I have been talking of the handyman who makes his own hutches, but if you wish, you can buy excellent hutches, both for indoor and outdoor use, in blocks of six or nine hutches, or singles whichever you prefer.

If you have a lawn or a grass patch, your cavies will be very happy to graze on it, so long as you have not sprayed it with weed-killer or fertilizer, and it has not been fouled by dogs.

You will require a run, which not only keeps the cavies in, but will keep cats and dogs out. It is also necessary to cover part of it to provide shade from the sun, and shelter in case of the unexpected shower. But do not put them out on wet grass. It is not necessary to have a netting wire base on the run, because cavies do not

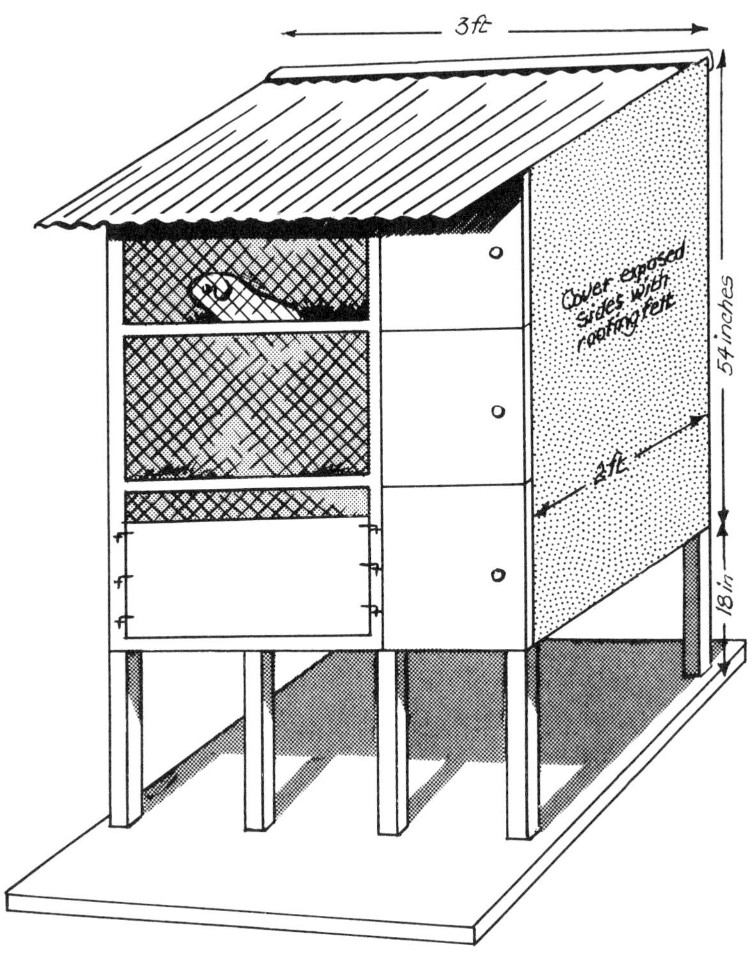

A Simple type of outdoor hutch, ideal for a sheltered corner of the garden. Corrugated plastic is ideal for roofing—on top of wooden hutch roof. The bottom hutch has a wooden shutter in position for night-time. The hutch is best standing on a concrete base.

burrow like rabbits but if you do have a wire base, it makes it easier to lift the run, and move it complete with cavies to a fresh area of grass.

FEEDING DISHES

Cavies very often throw their dishes around or use them as toilets, so the best feeding dishes should either be too heavy to throw around, or those which can be fixed. Among the best are the half-moon shaped dishes which hook on to the netting wire, and are deep enough to make it difficult for the cavies to soil them. They can be used as both feed and water dishes.

Water bottles such as can be bought in pet shops, are quite good if the cavies know how to use them, but often they chew off the end of the drinking tube leaving a jagged edge. Unfortunately, when I have enquired, I have been told that you cannot buy fresh tubes, but must buy the whole bottle.

BEDDING

Bedding in a hutch serves two purposes. It provides warmth and it soaks up urine etc., and so keeps the cavies clean. Some fanciers start with a layer of newspaper and, if you wish, you can sprinkle with a little disinfectant or dustbin powder—just a very small sprinkle. On top of this you need about two inches of clean wood-shavings. Sawdust is not satisfactory—if it is fine, the cavies inhale it, and it can cause a lung disease which is incurable. Peat can be used, but it can stain light-coloured cavies. Then add a good topping of hay. The cavies will eat the hay so it must be kept topped up—

ESSENTIAL FEEDERS AND DRINKERS

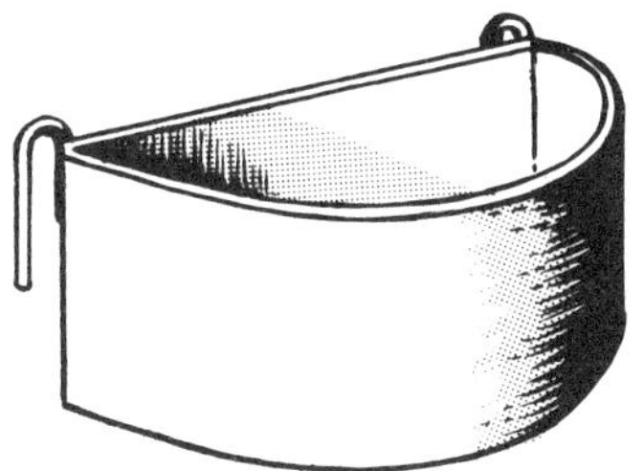

Half-moon type dish which hooks on the netting-wire.

You can set a dish in a cement base to prevent it from being tipped over.

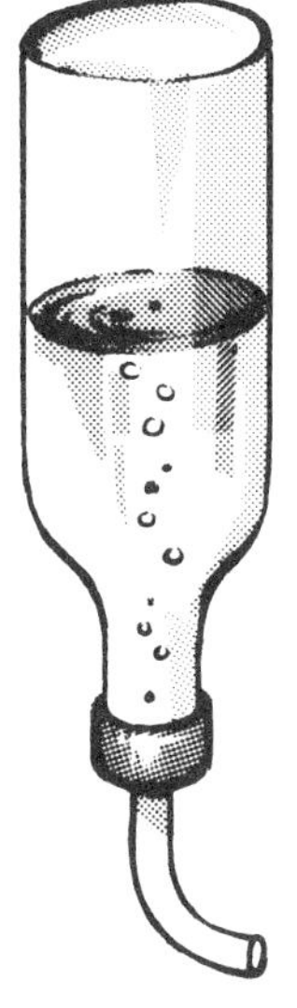

Drinking bottle which can be bought from a pet shop. Stainless steel or glass tubes are best—cavies soon bite through the aluminium ones.

Attach to the netting-wire with elastic band with the nozzle poking through.

twice daily is satisfactory. This can be done when feeding. Straw may be used in an emergency, but it must be soft. The hard, stalky type can do serious damage to the cavies' eyes, and straw is not such a good item of food as hay.

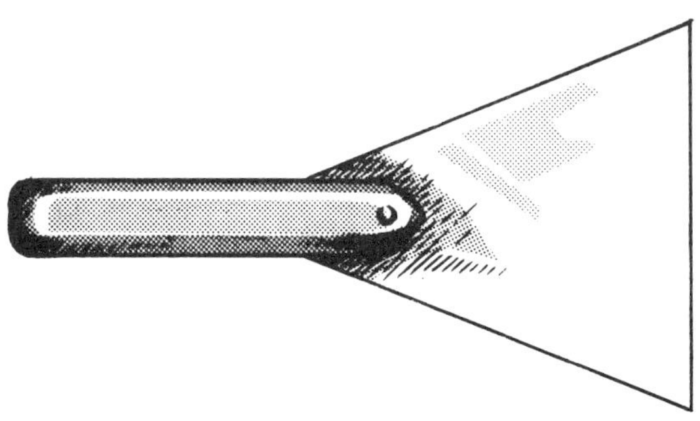

A paint scraper is useful for scraping the floors of the hutches when you are cleaning.

CLEANING

Clean your hutches at least once a week, unless the weather is very cold. Then, provided the bedding in the hutch is not wet on top, it can be covered with a fresh layer of shavings and hay. Of course you cannot go on doing this for long, but for a short time, during a very cold spell, it can help to keep the cavies warm.

Using a small shovel and scraper the hutches can be cleaned quickly. The soiled litter can be put into a bucket and then into a wheelbarrow, when it can be wheeled away to a suitable site for rotting to make garden compost.

CHAPTER 2

Feeding

ESTABLISH A DIET

When you buy cavies, it is a wise precaution to ask the vendor what his method of feeding has been. Then, if you wish to change the routine, do so gradually. The upset of moving to a new home is enough without also burdening your cavies with a change in their eating routine.

The essential items of a cavy's diet are:

> **oats in some form,**
>
> **greens or roots,**
>
> **hay and water,**
>
> **with supplements when necessary.**

The majority of fanciers give an oat mixture. I give a mixture which contains crushed oats, barley, flaked maize (cut out the barley and flaked maize in summer), rabbit pellets, ground nuts, sunflower seed, and puppy biscuits. A good handful for each animal daily is usually sufficient, but some cavies do eat more than others, so top up if their dish is empty. In winter, roots, such as carrots, beets, and swede are given with whatever garden greens are available. The thick stalks of Brussels sprouts, split down the middle so that the cavies can get at the marrow, are much appreciated. Beet pulp, soaked in water overnight and squeezed out and dried off with bran, makes a good mash which may be fed for a change instead of the roots.

Hay should be given ad lib, always giving more when it is eaten. Straw is not a good substitute, but if it has to be given on occasions, then use soft oat straw. It has never been established why hay is so important, but cavies deprived of it do not thrive, no matter how well fed they are otherwise.

FRESH WATER

Water is necessary, although many people maintain that they do not need it. It is thought that they get enough water from the greens and roots they eat, but can you be sure that they are getting enough greens and roots to give them the fluid they need? My cavies get greens *and* water. Some drink copiously; others drink a little. Be sure to wash out the drinking dishes quite regularly.

VALUE OF GREENSTUFF

When spring comes and the wild greens and grass are available, you can see the cavies brightening up. When you take a sackful into your shed, and they smell it, they get very excited. Grass is one of the best green foods for cavies. Lawn mowings are greatly appreciated, but they should not have been sprayed with any chemical, such as weed-killer, and do use mowings when they are *fresh*. If they have been lying in a heap and have become hot, do not give them to cavies.

When first feeding wild greens in the spring, feed a little and gradually increase it. By summer, you can fill the hutches, and the cavies will come to no harm. You will be surprised at the vast quantity they will eat. Mixed with the grass you can feed such weeds as sow-thistle, dandelion, plantains, dock, and cow parsley. But be careful *not* to feed poisonous greens such as hemlock, foxglove, buttercup, boxwood, bindweed, rhubarb, or potato leaves, etc. The small white clover is safe to feed, but do not feed the tall pink clover. Although such things as buttercups are poisonous when fresh, they are harmless when dried in hay.

SUPPLEMENTS

Cavies, monkeys and human beings are the only creatures who require to take in Vitamin C in their food. All other animals synthesise it in their bodies. In summer, if they are given plenty of greens, the cavies should get enough Vitamin C, but in the winter it is best to give some vitamin supplement. This can be given in the form of an ascorbic acid tablet daily in the drinking water. Rose-hip syrup or baby orange-juice are very good sources of Vitamin C, and can be given in the drinking water, or on the food. You cannot give an overdose of this Vitamin, as it is water-soluble, and any surplus will be excreted in the urine, but if you give cod-liver oil, you can overdose. As it is an oil, it is stored in the body and can cause paralysis. If you wish to give cod-liver oil, give a polyunsaturated oil such as "Vitapet".

Other supplements which are useful are mineral supplements such as "Stress" and "Vionate". Sprinkle these on the food as you might salt. "Bemax" sprinkled on the food is a good source of Vitamin B.

Glucose C. in the drinking water—at the ratio of two teaspoonfuls to a pint of water— helps to keep in-pig sows in good condition.

If in-pig or new-littered sows develop a sore on the back, I have found it helpful to give half a teaspoonful of soya flour mixed in a little milk, with a little bread soaked in it. Give this three times a week. This gives them extra protein. Do not give it to the in-pig sows unless they show signs of a break in the back, because it may increase the size of the babies, and make birth difficult. Nursing sows, and newly-weaned youngsters appreciate a dish of bread and milk in addition to the normal diet.

So, if you give your cavies oats mixture in the morning with hay and water, and either roots or greens in the evening, and top-up the hay, and give supplements when necessary, your cavies should keep very fit.

Motherless babies can be reared on 'Lactol', 'Complan' or S.M.A. baby food fed by droppers, for a few days until they lap on their own. It is better, however, to foster them to nursing sows.

If they have to be artificially fed, a little, fed often is best. Make the last feed as late as possible in the evening and the first one as early as possible in the morning.

CHAPTER 3

Breeding

SELECTING THE PARENTS

When you decide to start breeding, the first thing to do is to choose which boar to put with which sow. The boar should be the best you can get, as near to the standard as possible. Remember, the boar will pass on some of his good and bad points to the youngsters of every sow he mates. The sows pass on their points to their own offspring.

It is best when buying your first breeding stock to buy the boar and sows from the same breeder. Do not try out-crossing to another strain until you are more experienced. Many people think that in-breeding (mating together of near relations, such as mother to son, daughter to father, sister to brother) breeds weaklings, but this comes only if your breeding stock have these weaknesses in the first place, and they are intensified if both parents have them; so you can weed them out. If one of your breeding pair has a fault, choose a mate that excels in this point. If one or two of your youngsters are better than their parents, then you are on the right lines.

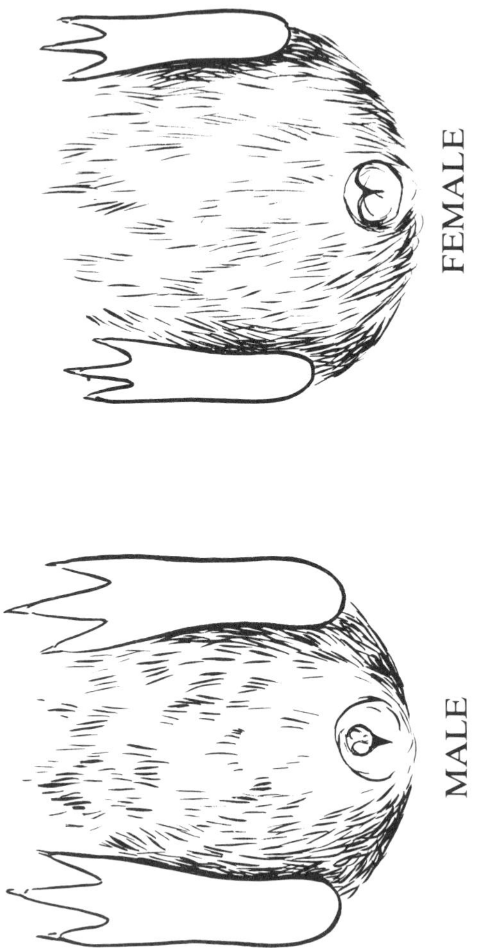

FEMALE
MALE
SEXING THE CAVY

SEXING THE CAVY

Many people, including many pet shop employees, find difficulty in sexing cavies. This is because, in the male, the sexual organs are carried within the body. Gentle pressure with a finger on either side of the orifice will cause the penis to protrude, and so make it easier to identify the male cavy.

In the female the orifice is Y shaped. The opening of the uterus is sealed by a membrane, which breaks every 16 days. This is the time when the sow will accept the boar. If she becomes pregnant, the membrane will not break again until the babies are born.

ESTABLISHING A STRAIN

When you have bred five or six generations without buying any outside stock, you can claim to have your own strain, but many breeders claim to have a strain even though they are continually buying breeding stock from other strains. If, however, you find that your stock all fail in any particular point, then buy outside stock which excels in this point, but watch carefully, because all the genetic make-up is not visible and faults may appear which were not obvious before.

PUTTING MALE AND FEMALE TOGETHER

Do not put young sows with the boar until they reach the weight of one and a half to two pounds. They should reach this weight by the time they are five months old. Do not breed with stock that is in any way unhealthy—or sows that are overweight. Stock in show condition can be overweight so put them in a big run to get plenty of exercise before breeding with them. Do not cut down on food. The weight of a full-grown sow should be three pounds or more. Peruvians and Shelties are usually heavier than average, and Himalayans less.

You can run more than one sow with a boar—even up to three or four—if you have large enough runs. The sow comes into season—that is when she will accept the boar—about once every two weeks and lasting a few hours. Occasionally, a sow just does not like a particular boar, and will not mate with him, so, in that case, you must put her with another boar. A sow should have her first litter before she is a year old. If she is older, there is a risk of difficult births, as the pelvic bones have set and do not open so easily to allow passage for the young. When the sow will not accept the boar of your choice, put her with another, even if it means a cross-bred litter—and do not listen to the old wives' tale that if they have a crossbred litter, this spoils them for future breeding of pure bred stock. This is nonsense!

Sows are usually left with the boars for several weeks, so it is not always possible to tell exactly when they have mated. The usual length of the pregnancy is sixty-five to seventy-two days—so, if you count sixty-five days from the day they are put together, that is the earliest date for the birth of the litter. Half-way through the pregnancy—at about a month to six weeks—you can feel the babies by placing your hands on either side of the cavy's abdomen. Always be very gentle when handling in-pig sows, and examine them only in a safe place where there is no danger of a fall if they make a sudden movement. To lift them from their hutches, use both hands, with your fingers under the belly to support it, and your thumbs over the back on either side of the cavy. You will often find that your sows get bigger and bigger, until you think they must have the babies or burst, but this is quite common and is because the babies are so well-developed at birth. It is not uncommon to have a litter whose *total* weight is more than half the weight of the mother.

About forty-eight hours before the birth, you can feel that the pelvic bones are opening, ready for the birth. Litters can number anything from one to six or more—three being the most common number. Cavies have only two nipples to feed their young, but can usually manage to raise four or five. With a large litter, the mother often loses condition, and needs a longer rest before being mated again. Many breeders reduce litter size by culling the weaker ones, but many—usually the ladies—do not like to do this; so it is a good idea, if one sow has a large litter, and another has only one or two, to let the sows share a hutch.

If the sow is still with the boar when the babies are born, he will not harm them deliberately, but the sow comes in season again an hour or two after the birth, and he will try to mate her; thus the babies may be injured in the scuffle. In any case, I think it is asking too much of a sow to be feeding one litter and carrying the next at the same time. If two in-pig sows share a hutch, and one of them litters, this may excite the other one so much that she may abort her babies, so it is best to put each sow on her own to have the babies, even though you may put two together while they are nursing.

DIFFICULT BIRTHS

One of the commonest causes of difficult births is very large babies, and there seems little we can do to control this. Sometimes, there is only one very large one, or it may be accompanied by several normal or small-sized ones. If the babies are three to four ounces at birth, that is quite big enough. If they are too small, they are often weaklings, and if too big they cause difficulties and are very often born dead. On one occasion, one of my sows—a Himalayan which are inclined to be smaller than average—after a difficult birth in which I had to assist her, produced two dead babies, one weighing eight ounces and the other eight and a half ounces. The mother herself weighed a little over two pounds!

What must you do, if one of your sows gets into difficulties? If the baby's head is protruding, you can grip it with a piece of clean cloth and gently ease it out. Be careful not to pull too hard or you may cause prolapse of the uterus. If the baby's head has not appeared and the mother is obviously in difficulties, you can insert your little finger and hook your fingernail on the baby's teeth, and then ease it out. If it is a "breach", sometimes a little circular-movement massage will bring it to the correct position. But if you are nervous about doing any of this, then your vet is the answer.

NORMAL BIRTHS

However, in 99 per cent of the births, the first thing you know of it is when you go to inspect your cavies and find one of them has a family running around. **It is always a source of wonder to see baby cavies, fully furred, eyes open, and running around within half to one hour of being born and nibbling grass at a day old.**

DATED PREGNANCIES

The usual procedure used by fanciers when breeding cavies, is to run sows with a boar, and remove each sow, when she is obviously "in-pig". The biggest drawback to this method, is that it is difficult to know when exactly the litters are due.

The entrance of the vagina of the cavy sow is covered by a membrane, which opens when she is "in season"— which is around every 16 days. If she mates normally at this time, then the membrane should not break again until the babies are born. Therefore one way of determining when litters are due, is to examine your sows daily, when they are running with boars. You will quickly be able to tell when the membrane has broken. Besides the open membrane there is redness and swelling of the nipples and the vaginal walls appear blue and swollen.

To examine a sow, hold her, resting on her back, along your left forearm, with her rump in your hand. Apply slight pressure, with first and second fingers of the right hand, on either side of the vaginal orifice, and you will very quickly be able to tell if she is in season. It takes only a few seconds. Make a note of the date when she is "in season", and continue to examine her daily. If the membrane has not broken again within 20 to 25 days, you can be fairly sure that she is "in-pig". It is usual to count 70 days from the date of the sow's last "season", to calculate the date of the arrival of the litter.

CHAPTER 4

Injury and Ailments

If cavies are well-housed, well fed, and not subjected to any undue stress, they are usually healthy creatures, but they do not stand up to bad treatment, such as keeping them outside in leaky, draughty hutches, being fed just occasionally when the owner remembers, or being mauled by small children. When you have an animal enclosed in a hutch—whether you are keeping a pet or breeding show stock, you take on responsibility for its very life. It is even more dependent on you for its health and well-being than are your own children. However, even in the best kept caviaries, accidents can happen, or illnesses occur.

In the past, the idea seemed to have been that a "sick cavy is a dead cavy" but gradually as more is learnt of the illnesses which affect cavies, we find that at least some of these illnesses can be successfully treated. True, a cavy appears to lose the will to live very quickly, and sometimes they die before you have the chance to find out what the trouble is, but we should all have a little knowledge of the common complaints, and what can be done about them.

BROKEN TEETH

The most common injury which a cavy can sustain, if it falls or is dropped, is broken teeth—often accompanied by a cut lip. The cut will usually heal without trouble, but with broken teeth, the cavy will not be able to chew his food properly, will quickly loose weight and will become emaciated. So give him "slops", such as bread and milk, or soaked pellets, with the addition of bran, for roughage, mineral and vitamin supplement. Grated carrots may also be given. The teeth will grow again. Cavies' teeth are growing all the time, but the chewing and grinding against the opposite teeth acts as a file which keeps them at normal length—but when some teeth are broken or missing, the corresponding teeth in the other jaw just keep growing and can reach enormous lengths if nothing is done. In such a case, the overgrown teeth can be cut with nail clippers and filed. You will be surprised at how easily this is done. Get someone to hold the cavy for you while you do it, and be very careful not to cut the tongue. Cutting the teeth is not painful to the cavy, but he will not like to be kept still and may wriggle. The teeth must be cut periodically until such time as the opposing teeth grow. Back teeth may sometimes be overgrown and cause distress. It is not easy to look at the back teeth, but if your cavy is continually slobbering, this may be the trouble. It is usually caused by a congenital deformity, and it is necessary to get a vet to deal with it. The chances are that they will grow again in the same way, so the prognosis is not good.

BROKEN LEGS

If a cavy has a leg broken, it will heal if it can be set and secured in position by plaster of paris. I once had a cavy sow which had its leg trapped. When I found her, the leg was nearly severed and had splinters of bone protruding. My vet set it in plaster of paris which was removed ten days later. Although the leg was deformed, it had healed, and the sow—which we renamed "Pegotty"—lived long and bred many nice babies.

FIGHTING INJURIES

An adult boar will usually happily share a hutch with a baby boar, but never put two adult boars together, or you may have quite a serious fight on your hands. Sows also may fight, when establishing hutch supremacy, but this is usually not so serious. Only very occasionally you may have a disagreeable sow, who makes the lives of her hutch-mates a misery. She is best put with a boar—or on her own. However, boars fighting can often result in injuries, such as torn ears, which usually heal without trouble, but split ears spoil a cavy for show purposes. While the edges are still raw, stick them together with little strips of adhesive tape. Sometimes this makes the edges join together again. Bites on other parts of the body should be washed with weak T.C.P. and dabbed with acriflavine. Most cavies will not bite human beings deliberately, but be very careful if you are trying to separate fighting boars or you may be badly bitten yourself. Put on a thick glove or wrap your hand in a towel.

EYE INJURIES

It sometimes happens that a cavy has a stalk from the hay in his eye, which causes the eye to become opaque. This usually clears in time. A little medicinal liquid paraffin, dropped in the eye, soothes it, or Golden Eye Ointment or Albucid Drops. Sometimes a grain of oats becomes lodged in the eye, and must be removed with a moistened piece of cotton wool.

SCABS AROUND MOUTH

Occasionally a germ may enter a cavy's cut lip and cause sores around the mouth. These are often difficult to cure. One of the best treatments is to paint it with Gentian Violet—obtainable from chemists.

LYMPHADENITIS

If you are not very careful, when feeding hay and greens, to ensure that it is free from thistles, your cavy may get a thistle thorn embedded in the throat. This sets up an inflamation which can lead to a large swelling on the throat. This rarely bursts of its own accord and usually has to be opened with a scalpel or sterile razor blade. Be sure your cavy is sitting on old newspapers which must be burnt later, as vile-smelling pus will pour from the swelling when it is opened, which should then be washed with weak T.C.P. solution. It usually fills up again, and has to be reopened, probably about a week later, and then it usually clears up. It is hopeless trying to bandage it; the cavy will have the bandage off somehow, so keep him in a hutch on his own, with plenty of soft hay. If the wound is discharging, change the hay daily, and burn the dirty hay. Afterwards disinfect the hutch thoroughly.

ABSCESSES

Very occasionally a cavy will develop a boil or abscess; these usually come to a head and burst. However, you can bring them to a head more quickly by smearing with warm mag. sulph paste. Occasionally you may have to open an abscess, but be sure it has reached a head first.

SKIN COMPLAINTS

The cavy seems to be particularly prone to skin trouble from many different causes. One of the commonest skin complaints is called all sorts of names from "Sellnick" to "Rat-mange". It is caused by a mite which burrows under the skin, the mite itself not being visible to the human eye. The cavy first of all shows tiny, raised spots on the skin which becomes scurfy and the hair falls out. The skin itches and so the cavy scratches, so if your cavies are losing their hair and have scratch marks on the skin, this is probably what it is. There are several different treatments—the most effective being to dip in a solution of "Tetmosol"—an I.C.I. product obtainable from Boots. The strength of the solution should be one part of "Tetmosol" to seventy-two parts of warm water. Just dip the cavy in the solution, hold for a few seconds and quickly immerse over the head; then dab off the surplus moisture with a towel, and allow to dry naturally. With a bowl of the solution, you can dip all your cavies quickly. Repeat the process twice more at fortnightly intervals. Many fanciers dip their cavies

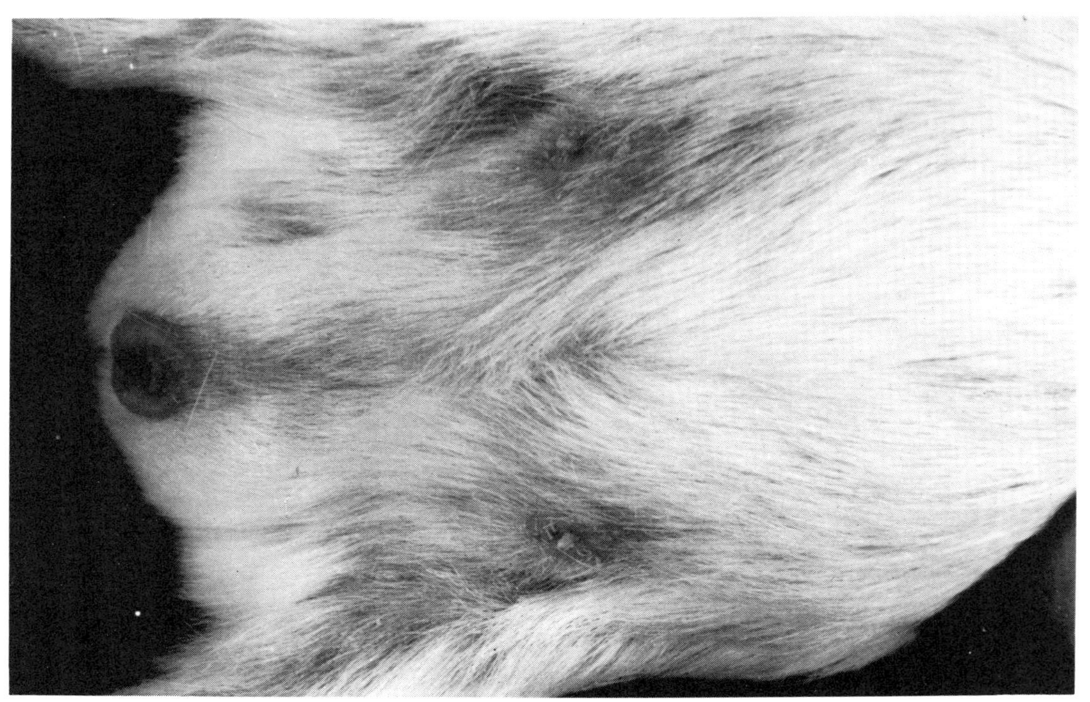

Figure 1 Sellnick Mite Infestation (Early Stages)
(Courtesy: Prof. H. Elvidge, Nuffield Institute for Medical Research, Oxford)

periodically as a preventative measure. It has never been established where the mite comes from, but the No.1 suspect is hay! Other treatments include grooming with paraffin, shampooing with insecticidal shampoos, and even a weak solution of sheep-dip.

LICE

Some of the treatments listed under "Skin Complaints" are also effective when dealing with lice. These usually inhabit the back of the cavy and can be seen as very tiny moving specks. They appear to do little harm, but nevertheless must be eliminated. Any of the insecticidal sprays or powders available at the pet shops may be used, but I would not use them on small babies. However, whatever you buy, do read the instructions carefully before use.

SCURVY

Many people think that "scurvy" means "full of scurf", but this is not so. Scurvy is a vitamin-deficiency disease, due to lack of vitamin C. It is characterised by severe weight loss, weakness, bleeding from gums and under the skin, and loss of hair. A few doses of vitamin C in the form of ascorbic acid, rose-hip syrup, or baby orange juice, and plenty of fresh greens, and you will see a marked improvement.

POST-NATAL SORES

Sows, after giving birth to a litter, may develop a sore on the back. This is often due to a mineral-protein deficiency. In late pregnancy, give sows about a quarter

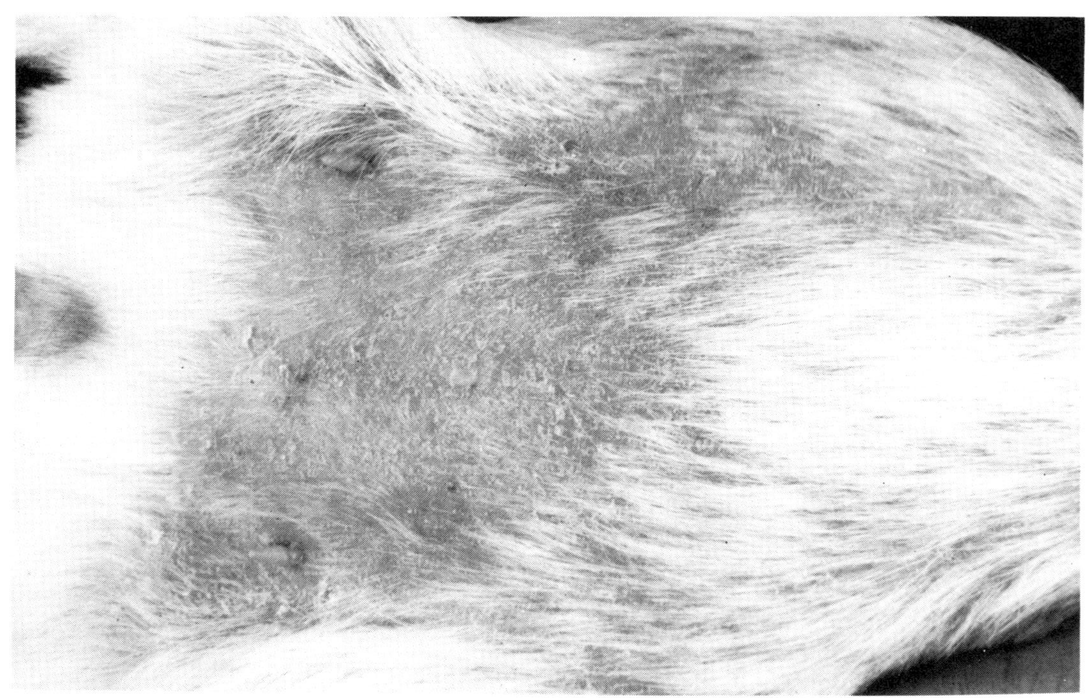

Figure 2 Sellnick Mite Infestation (Advanced State)
(Courtesy: Prof. H. Elvidge, Nuffield Institute for Medical Research, Oxford)

32

of a teaspoon of soya flour mixed with a little milk stirred into their food, and sprinkle with "Bemax" and mineral salts. You may find it beneficial to give fish liver oil in very small quantities—just a few drops. "Vitapet" is one of the more refined polyunsaturated oils which is better to use than unrefined Cod Liver Oil.

Occasionally, after they have littered, you will find sows with raw areas on the rump and under the belly. This is caused by the sow pulling the hair out and biting herself while she is cleaning up. It could be that she had a prolonged labour and difficult birth, and was too tired to clean up immediately. Then as the blood, etc., dried on her she had difficulty in removing it. If you notice in time, then it is possible to sponge her clean and dab her dry; this may then stop her from doing any damage. If the damage is already done, then dab with a little "Vitapet" or castor oil. If it is very bad, apply some gentian-violet or acriflavine. Broken skin not connected with littering are usually the result of too much of certain cereals in the diet—such as flaked maize and barely—in hot weather. So try to cut these out during the summer and give more pellets. A touch of cod-liver oil on the affected parts helps.

PARALYSIS OF HINDQUARTERS

This is usually caused by an injury to the spinal column which could come about by something as simple as a sudden jump by a startled cavy. If it shows no sign of improving in a couple of days, the cavy should be put down. Sometimes, too much cod liver oil has the effect of paralysing the back legs. If you are giving too much at a time, it will accumulate in the body, so omit it and see if the cavy improves. If you resume the cod liver oil, give much smaller doses, and ocasionally stop it for a period. Here again, I would recommend the more refined "Vitapet" oil.

IMPACTED RECTUM

As cavies age, especially boars, you may find they have a lump of faecal matter in the rectum. The only thing you can do is to squeeze it out gently with soft tissue and clean the area with olive oil. Be prepared for a most unpleasant smell. I am afraid, in most cases, you will have the job of clearing it out weekly for the rest of the cavy's life, although if done regularly, it will not smell so badly.

RESPIRATORY DISEASES

Snuffling in cavies can be caused by several different things—a virus infection such as the common cold, or dust such as fine sawdust. This can cause a lung condition which is chronic, so use shavings rather than sawdust for bedding. Another cause can be in the breeding. Some babies with short faces and round heads appear to develop snuffles very easily and often die young, so it would appear that there may be some defect in the nasal passages due to the shape of the head. A little "Vick" or camphorated oil—or similar preparation—just dabbed on the sides of the nose often helps, or a small dose—about one-eighth of the recommended dosage—of a children's cough mixture. When the trouble spreads to the lungs, the cavy develops pneumonia, and this is a serious killer disease. Sometimes there is a rust-coloured discharge from the nose, and the cavy rattles as he breathes. They can be treated with sulphamethazine or tetracycline by mouth. Be sure the cavy takes plenty of fluid, even if he should be off his food. If he will not drink, give fluids by dropper. Give glucose and water for a few days, then a little milk and fresh green food if he will take it. Be sure you isolate him, as pneumonia can be infectious and can spread quickly to other cavies.

SCOURS

"Scours" just means diarrhoea, and it may just be a tummy upset caused by eating something which disagrees with it. Whether he will recover or not depends on what it was and how much he ate. A cavy cannot vomit, so whatever he eats must go through. Half a teaspoon of liquid paraffin helps to clear it out and soothes inflamed intestines. However, if you find that your cavies are scouring and dying, and that there are fresh cases every day, the chances are that it is a salmonella infection, which is a real killer, and can wipe out all your stock in a very short time. Treatment is of little use. The cavies usually die within twenty-four hours, and if they do recover, they may be carriers and infect other cavies later. The only practical thing to do, is to isolate your *healthy* cavies. Take them away from the sick ones and place in another building, putting them in cardboard boxes if you do not have another supply of hutches. Your caviary is already contaminated, so if you move the sick ones, you not only contaminate other premises, but you also leave the healthy cavies where germs are already swarming. Of course, if any more become ill, they must immediately be taken out. Each day feed and attend to the healthy ones before attending to the sick ones, and have a good wash afterwards. Burn any dead cavies and their bedding, and when it is all over, thoroughly disinfect your caviary and hutches; then leave for several weeks before you bring fresh stock in. It is a good idea to burn hutches, but these are so expensive nowadays that that might mean giving up altogether. Wash them with a strong solution of soda, and fumigate your shed. With a disease like this, prevention is the all-important thing. The disease can be brought in by mice and rats, so keep your premises free of them. It can be spread from hutch to hutch by flies, so try to keep clear of those. Cleanliness of feeding and water dishes, hutches and premises is also important. "Vanodine" in the drinking water—enough to turn the water pale yellow—acts as an intestinal disinfectant. "Vanodine" can also be used to wash utensils and spray round hutches and shed.

PREGNANCY TOXAEMIA

This is so called because it occurs mainly in sows in late pregnancy, but occasionally it may occur in other stock. An overweight sow, in late pregnancy, is close to physiological breakdown, and any extra stress can bring on toxaemia. In tests made by J. R. Gannaway and A. M. Allen in 1971, all that was necessary to induce toxaemia was to withhold supplemental green food. When the cavy becomes stressed, the liver overacts and vast quantities of stored fat enter the blood-stream at poison levels. The blood becomes too acid; the kidneys work overtime trying to clear it, and the cavy dies of kidney failure. The cavy usually appear to have twitching of the muscles, and goes into a coma and dies. Little can be done at this stage, but if you make your cavies fit but not fat before breeding with them, you may never encounter this complaint.

CONCLUSION

There are of course many other complaints, but I have outlined the ones which you are most likely to encounter, and also a number that you can prevent.

One word of warning—while it is reasonable to give very small doses of children's cough mixture to cavies— not all medicines suit cavies. Penicillin causes anaphylactic shock and usually kills them—even though it can safely be given to a mouse.

CHAPTER 5

The Self Cavies

The word "self" when used to describe colour means "the same colour all over", but it is also the name of a variety of cavy. The Self variety, are, as you would imagine, the same colour all over, but they are also smooth, short coated; therefore you *cannot* show, for example, a self-coloured Abyssinian in the Self classes.

LOOKING FOR THE "PERFECT" SPECIMEN

There are thirty points for **colour** in the Self Standard, but there are nearly as many—twenty-five—for "type". Type can be described as the shape of the cavy. **It should have a broad head, a short face with plenty of space between the eyes, and broad high shoulders. Eyes should be large and ears should be petal-shaped and drooping.**

You will find that the poor type, snipey nosed selfs tend to have smaller eyes and the narrow face gives less width between eyes. Young cavies are apt to go a bit "rangy" and long while they are growing up, but notice the shape of your babies when they are a few days old. That is when you see the shape they will have as adults, so do not be disappointed if they appear to go long in head and body for a bit, while they are growing.

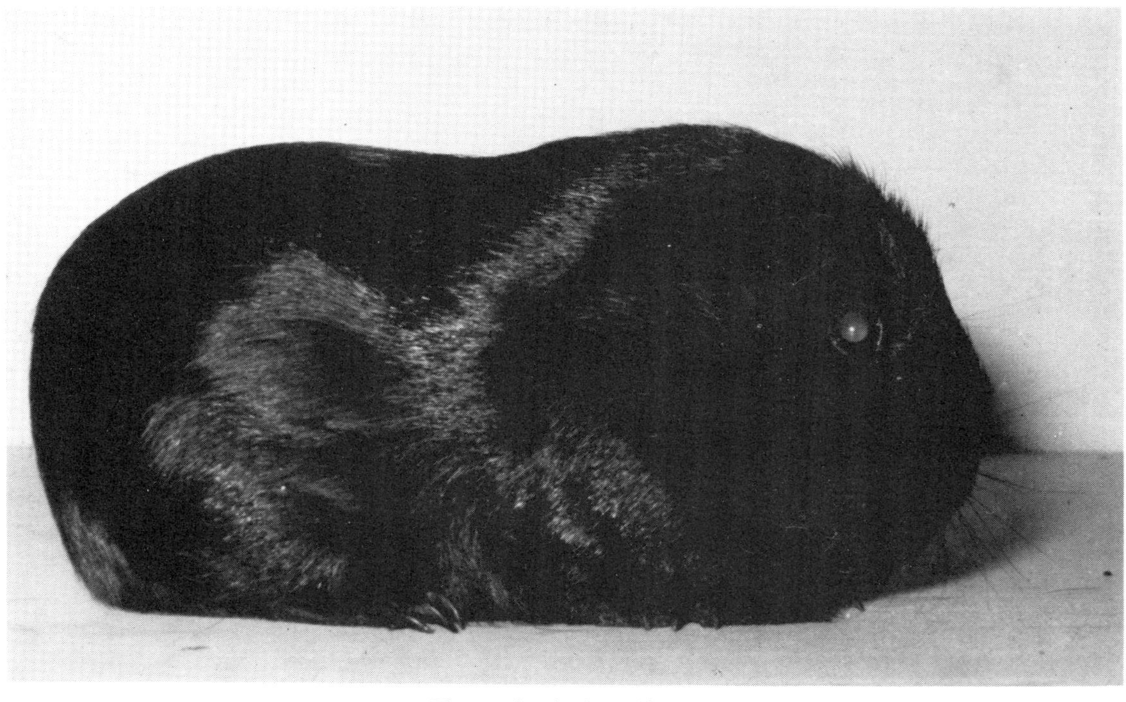

Figure 3 Black Self Cavy
(Owned by Mr. George Howard)

GROOMING AND TRAINING

The coat should be short and silky, and grooming is very important, to be done regularly and not just before a show. Groom out the long guard hairs by drawing your thumb and forefinger along them; then rub from head to tail with your hands. A rub with a piece of pure silk will give them that extra shine. Besides making them look well, grooming gets them used to being handled, and they will sit still while being judged. A Self does not show itself to advantage if it is scurrying round the table, or stretching itself out to see what is going on.

Self sows are better than boars for showing. Boars tend to have harsher coats and not such well-shaped heads as sows. That does not mean, however, that there is never an outstanding boar—far from it!

RECOGNIZED COLOURS

The Selfs are of the following colours:—

Black

This is the most popular colour, and a good black, well-groomed and shining like a top hat, is a beautiful animal. Common faults are red or white hairs, poor undercolour—the jet-black colour should go right down to the skin—poor belly colour and poor type. You also see some with small or pricked-up ears and small eyes. However, there are so many good blacks that competition is quite fierce, and blacks are very often the "Best-in-Show" winners.

Red-eyed Whites

Are also very popular, and good specimens in this colour usually excel in type. Whites are not plagued with

Figure 4 Champion Golden Self "Showers of Gold"
(Owned by Mr. George Howard)

hairs of another colour, nor can they have poor under-colour, but they are difficult to keep in the spotless condition required for show-standard whites. Any staining, especially on the belly, can be difficult or impossible to clean.

Dark-eyed Whites

Should be the same as Red-eyed Whites, except for the eye-colour, but they tend to have some pigmentation of feet and ears. When breeders succeed in breeding out this fault, they find that the eyes lose pigmentation as well, so this makes them a difficult variety to breed. Nevertheless, some good dark-eyed whites have been shown though they are few and far between.

Creams

The cream Selfs have generally very good type. Where they fail is in undercolour. The desired shade of cream is a pale delicate cream, which inevitably has a very pale—almost white—undercolour. Unfortunately this very often makes it lose points when competing against other colours. In litters of young creams it will surprise you to find such a variety of shades. The pale ones are the ones you want to show. A white boar can be used to breed with creams, but you will get a number of white youngsters. The eyes of the cream are deep ruby which look black in certain lights.

Goldens

There is always some controversy about what is the desired shade of golden. It should be a rich golden shade often described by new fanciers as "ginger". There should be no suggestion of yellow. Many goldens fail in undercolour, but it is always a joy, when judging, to find those with as rich a colour against the skin as on top. If the undercolour is too pale, it gives the appearance of flakiness—streaks of lighter colour—especially on the sides at the rump. The eyes of the Self Golden are pink. The dark-eyed Golden is dealt with in the chapter on the Rare Varieties.

Reds

These should be a rich dark mahogany colour with ears and feet to match. Eyes should be dark ruby. Young reds are apt to get lighter as they grow older, so the ones to keep are the very dark youngsters. Strangely, if you have a sow with a small patch of white for example a white foot, she is very likely the one that will breed the darkest babies. Some youngsters may have white hairs. These usually clear, and leave a good rich red colour. A number of good Reds are being shown. In the past they tended to be rather small and snipey, but they are now vastly improved.

Chocolates

These should be a rich dark chocolate—*not* a milk-chocolate colour—with eyes to match. These have improved greatly in recent years. The biggest fault is poor colour—or fading colour on the belly with golden or white hairs. A few years ago, this type was not good, but as indicated, this has improved considerably.

Beige

To those who know a little about genetics, the beige is the pink-eyed dilute of chocolate. We are told that the beige should be the colour of beige cloth, but with so many shades of beige cloth, this is no real guide. The best way of describing it is a "very pale beige with a pink tinge". In a litter of young beige, you will find many shades from very dark to very light. Keep the dark sows for breeding, but use a boar of a light colour or vice versa. Beige cavies usually excel in undercolour. The lovely pinky beige colour goes right to the skin. One fault to watch is bars of lighter or darker colour making the colour uneven. Type is not usually as good as in blacks and whites.

Lilac

As the beige is the pink-eyed dilute of chocolate, the lilac is the pink-eyed dilute of black. The colour to aim for is pale dove grey with a pink tinge. It is not an easy colour to breed, and is apt to have dark bands. Eyes should be pink.

Saffron

This is the pink-eyed dilute of cream dealt with in the Rare Varieties sections. It should not be shown in self classes, but in Rare Variety or unstandardised classes.

PENALTIES

The *Standards* which follow show the maximum points to be awarded for each important attribute. Deductions will be made for any imperfections; e.g. white or red hairs in Self Blacks.

GUIDE FOR JUDGING

Standard for all Self Cavies

Colour: .....Deep and Rich 30
Shape:......Short, Cobby Body:
Deep Broad Shoulders
Roman Nose 25
Coat:.......Short and Silky 15
Ears:.......Rose-shaped, set slightly
drooping: Good width
between 10
Eyes:.......Large and Bold..................10
Condition: 10

———

Total 100

Remarks

Blacks: Colour should be deep and lustrous, the same going down to the skin. Eyes black.

Chocolates: Should be rich, dark colour, with ears and feet to match. Eyes Ruby.

Reds: Should be rich deep colour, with feet and ears to match body. Eyes Ruby.

Creams: Should be pale, even colour with under-colour to match, and free from lemon or yellow tinge. Eyes ruby.

Whites: Should be pure snow-white throughout; ears and feet to match body. Eyes pink or dark ruby.

Beige: Should resemble real beige cloth, with ears and feet to match body. Eyes pink.

Goldens: Should be a rich golden shade, with no suggestion of yellow. Ears, feet and toe-nails to match body. Eyes pink.

Lilacs: Colour to be pale lilac carried down to skin. Eyes pink.

CHAPTER 6

Agouti Cavies

The Agouti cavies have a ticked or speckled appearance but—unlike roans and brindles, which have an even intermingling of two different colours, with each hair completely of one colour—the Agoutis have bands of two different colours on the hairs interspersed with plain-coloured hairs.

The most popular are golden and silver, closely followed by cinnamon, but many new colours are developing, and old colours which had disappeared are being brought back. In America, there are Agoutis which have the ticking on the belly, and are known as "solids", but in Great Britain, show Agoutis have a plain belly with a different base colour. The belly appears to be of one colour because of the absence of the interspersing contrasting hairs, but if you brush the belly hair back with your finger, you will see the contrast at the base of the hairs.

PRINCIPAL COLOURS

Golden Agouti

Base colour, black or slate, ticked with golden and with golden-red belly colour.

Silver Agouti

Base colour blue-black, ticked with silver: Dark silvery belly colour.

Cinnamon Agouti

Base colour brown-chocolate with silver ticking and dark cinnamon belly colour.

Lemon Agouti

Black base, cream ticking.

Orange Agouti

Like a golden agouti, but with a brown base.

Chocolate Agouti

Chocolate base, gold ticking.

It is possible to produce agoutis in dilute colours such as beige/golden, lilac/golden and so on. These are pink–eyed. The banding on these varieties is not as clear because of the dilution, which sometimes makes fanciers question whether or not they are agoutis.

These pink–eyed agoutis are at present known as Argentes, and there is more about them in the Rare Varieties Guide Standard Section. The exception is the salmon agouti; although at present extinct, it is accepted by the Agouti Cavy Club. It is salmon–coloured with white ticking.

FAULTS

Long coarse guard hairs which spoil the appearance of the ticking should be groomed out. Uneven grooming leaves dark patches. Feet should be ticked to match the body. Often they are too dark. The chest and under the chin should also match the body. They may be too light. Light bands under the chin are called "bonnet strings" in show reports.

Belly should be narrow and have a clear demarcation between body and belly colour.

Eye circles—that is a circle of light-colour round the eyes is yet another fault.

STANDARDS OF EXCELLENCE

The breakdown shown below indicates the maximum points which may be awarded.

Standard for all Agouti Cavies

    Colour 20
    Shape 20
    Evenness of ticking 30
    Eyes, large and bold 5
    Ears, well-shaped and slightly drooping . . . 5
    Size combined with quality 5
    Coat and Condition 15
                                                    ———
    Total 100

Remarks

Goldens   should be of a rich golden hue, with even, dark ticking throughout, including feet and chest. The undercolour must be a deep blue-black carried well down to the skin. All four feet must be evenly ticked and free from brassiness—feet which are too dark must be avoided. The belly colour should be a deep rich golden, free from brassiness and as narrow as possible.

Silvers   should have a rich black undercolour with even silver ticking throughout, including chest and feet, the belly colour of a rich dark appearance and as narrow as possible.

Cinnamons should have a deep cinnamon undercolour with even and sparkling silver ticking throughout, including chest and feet. The belly colour should be of a rich dark appearance and as narrow as possible.

Faults to be penalised

Eye circles, broken coat, side whiskers, excessive white hairs and odd feet.

ILLUSTRATION

A photograph of a prize winning Agouti is shown on plate 4 of the coloured plates.

CHAPTER 7

The Abyssinian Cavy

The Abyssinian is the rosetted, rough-coated cavy which is often confused with the Peruvian by the general public at shows. Many times, we hear people say, "Oh! There is a Peruvian!" In fact, the Abyssinian's coat should not exceed one and a half inches in length. It is because the coat does not lie flat on the body that people receive the impression of a long-coat. **The rosettes and ridges, formed by the unusual way in which the hair grows, are the important points in an Abyssinian.** It should have four rosettes in a straight line over the saddle of the body, four around the rump, and one or two on each shoulder. Where the hair growing in one direction from one rosette, meets the hair growing in the opposite direction from another rosette, it forms a ridge. The rosettes should be so placed that the ridges run in straight lines, both across the body and down the back and sides. For example, if the saddle rossettes are out of line, this would cause unevenness of the ridges. The rosettes should be deep with the hairs radiating from a pin-point centre. The coat should be harsh, and there should be no flat coat anywhere. This includes the head. The Abyssinian has the appearance of having a

Figure 5 Abyssinian Cavy
(Owned by Mr. George Howard) Note the position of the rosettes (see text)

moustache.

COLOUR NOT IMPORTANT

Colour is unimportant. They can be any colour or any combination of colours. Among the favourites are roan, brindle, and tortoiseshell and white, and there are other colour combinations. Unlike the smooth-coated varieties, the Abyssinian can be marked in any way. Self Abyssinians in various colours are also shown, but it is often found that these lack the necessary harshness of coat. White Abyssinians are not very successful as show animals. Not only are they soft in the coat, but often they are not very robust and lack condition. Attempts have been made from time to time to breed Himalayan-marked Abyssinians; there is no great difficulty in achieving the markings, but, once again, there is the problem of the soft coat.

FAULTS

Breeding good Abyssinians is not easy, but as soon as the mother has cleaned them up after they are born, you can see the faults—and good points.

Faults to look out for are:

Out of line, or double-centred rosettes, or rosettes with an open or elongated centre: uneven ridges: not enough density, or length of coat, making the rosettes look flat: missing rosettes or double rosettes. Then you must look for soft coats, although you must make allowances for youngsters' coats are usually much softer than adults.

Abyssinian boars have a reputation as fighters, not

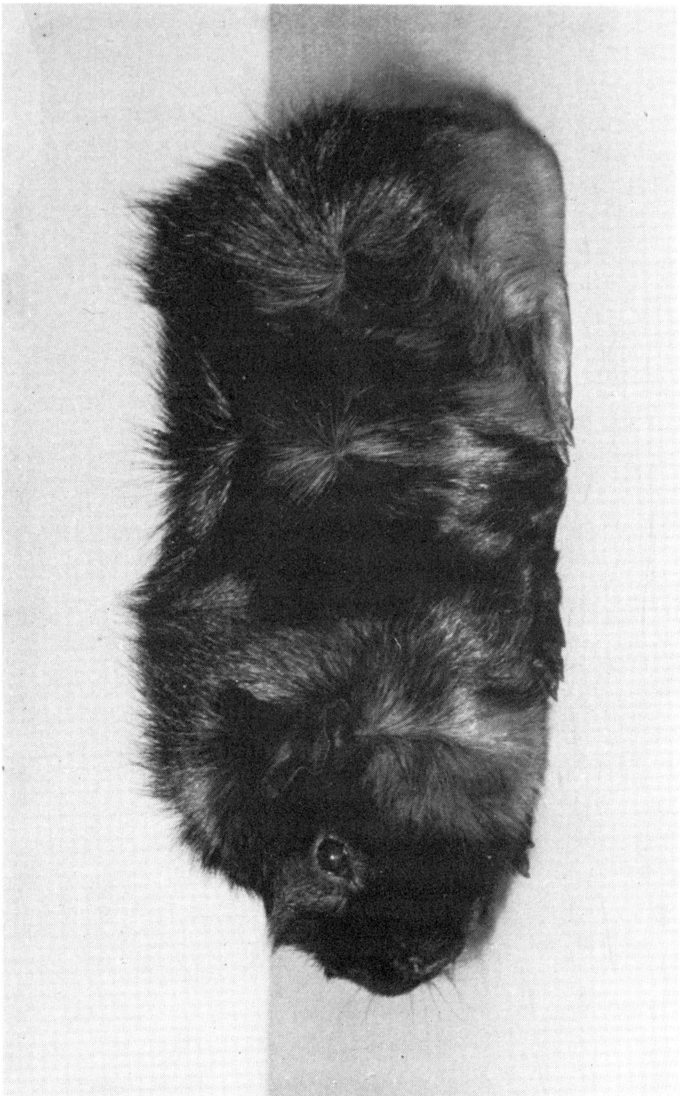

Figure 6 Abyssinian Cavy
(Owned by Mr. George Howard)

always, but sometimes deserved, and as Abyssinian boars are usually better show animals than the sows, very often there is a line-up of boars on the judges' table. It is wise to treat these boars with respect, but firmly, if you are stewarding—especially if they start grinding their teeth at each other, a sign that they are angry.

GROOMING

The best way to groom an Abyssinian is with a toothbrush, brushing outwards from the centre of the rosette. It is often said that the Abyssinian is the lazy man's cavy because it needs less show preparation than any other variety, but that "little bit of grooming" can mean quite a difference. Again bathe them several days before a show, so giving the coat time to regain its harshness. A dip in rain-water is often considered to be of advantage.

Standard for the Abyssinian

Rosettes	20
Ridges	20
Coat	20
Shape and size	10
Head furnishing and mane	15
Colour	5
Eyes and ears	5
Condition	5
Total	100

Definition of Colours

Tortoiseshell
and White Red, Black and White colours separately placed in patches; no definite order of patches laid down. Clearness of colour adds to attractiveness and an advantage.

Tortoiseshell Red and Black. Patches as clear as possible.

Brindle Red and Black hairs interspersed. If red hairs predominate the animal is a light brindle: if dark hairs predominate, the animal is a dark brindle.

Note: Occasional quality specimens of the above appear with a white foot or white toes. It was officially decided at an Abyssinian Cavy Club General Meeting that, providing the white area did not exceed the size of a shilling (or 5p. piece) it is in order to exhibit same in Tortoiseshell or Brindle classes.

Selfs Blacks, Reds, Whites: other colours rare.

Roans Blue roan is slate grey and interspersed with white hairs. Strawberry roan is red hair interspersed with white hairs.

Off-colours Not classified in the aforementioned categories—Red and white: Black and white, Agoutis, Himalayans and occasional various colours.

Plate 1   *Top*   Sheltie Cavy (Mrs Pat Wood).
           Bottom   Peruvian Cavies (Mrs I. Routledge).

CHAPTER 8

The Dutch Cavy

Although the Dutch variety is fairly popular, it is so difficult to breed a really good one that there are relatively few of them to be seen at shows. The Dutch is basically a white cavy with coloured markings. The commonest colour is red, followed by black; but there are also chocolate, agouti, yellow and cream Dutch.

The rear end of the body should be coloured, and the front white: the demarcation is a straight line round the body as near to the front legs as possible without the colour being present on the legs themselves. This is called the "saddle".

The hind legs should have white socks of equal length going half-way up to the first joint, and having a straight demarcation line round the foot. These socks are known as "stops".

MARKINGS OF PRIME IMPORTANCE

The coloured markings on the face should be as round as possible, including the eye and ear, reaching the whisker-bed, but not actually going into it. The colour should not run on to the neck or under the chin, and the markings on either side of the face should be separated

Figure 7 Sextuple Champion Dutch Cavy "Pandora"
(Owned by Mr. George Howard)

by a wedge-shaped blaze of white, extending up the centre of the face to between the ears.

The markings on either side of head and body should be identical, giving a well-balanced appearance.

This, of course, is the ideal, and you can imagine it is a difficult one to achieve, so much so that even the best Dutch are not without faults; so do not be disheartened if yours are faulty. They may still be near enough to the standard to make it worth showing them.

COMMON FAULTS

Here are some of the common faults: the saddle may look straight on top, but it may be very uneven on the belly. Cheek markings frequently drag under the chin or down the neck, or they may go right down the whisker-bed to the cavy's mouth. On the other hand they may be too high, finishing just under the eye, and, of course, it is quite possible that one cheek is too high and the other too low!

Flesh-coloured marks on the ears constitutes something else to avoid. The ears should be coloured inside and out. Eyes should match the dark portions of the body. Short stops are a common fault, some of them barely covering the cavy's toes; or they may be too long extending too far up the leg. Again, they may be unbalanced with one long and one short stop.

COLOURS IN EXISTENCE

While markings are more important than colour in the Dutch cavy, colour is of importance too, and they should be as near to the standard of each colour. One colour you will find different is the red; this is not usually as dark as a red self, but more like the golden self colour, but richness of colour is important. The black Dutch is eye-catching with its black and white pattern. Look out for white hairs or even red hairs on the black parts of the cavy. Sometimes the red hairs disappear as the cavy grows older, and often these are the ones with the best depth of colour.

The silver agouti Dutch has, in the past, been difficult to breed. It was found that often the youngsters were weakly and difficult to rear, and many breeders gave up because of this. There are still very few to-day, and they are usually not so well marked generally as the other colour. Golden agouti Dutch are fairly popular: the latest addition to the Agouti Dutch to be recognised by the Dutch Cavy Club is the cinnamon.

There are a few cream Dutch around, but they are rare. The biggest drawback to this colour is the fact that the demarcation line on the belly is not clear. The cream colour appears to blend into the white.

However, these rarer colours do have their dedicated breeders, so we may see more of them in the future.

FAULTS IN A DUTCH CAVY

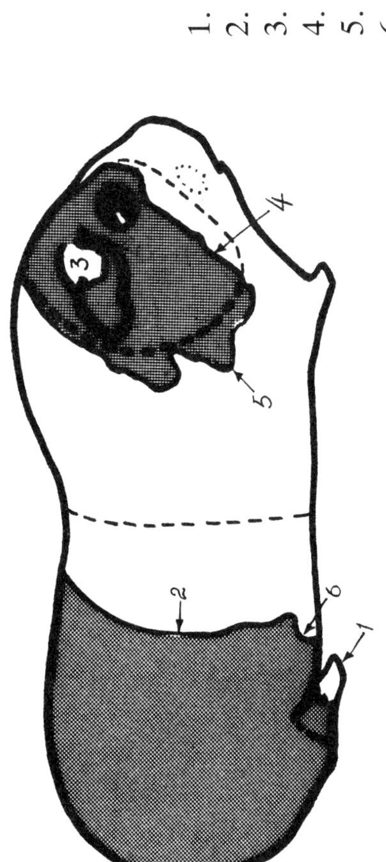

The Dutch Standard

Blaze and cheeks 15
Clean neck 10
Saddle 10
Undercut 10
Feet stops 15
Ears 15
Eyes 5
Colour 10
Size, Shape and Condition 10
                                            ——
Total 100

CHAPTER 9

The Tortoiseshell and White Cavy

The Tortoiseshell and White, or *Tort and White* as it is usually called, is considered to be the most beautiful and the most heart-breaking of all the cavy varieties. Heart-breaking, because you might breed a hundred—or more—without breeding one anywhere near show standard no matter how high the quality of your breeding stock. A good one, if it should be offered for sale, could cost a very high price, more than any other variety, but it is very rare that a breeder will part with a top-class animal, so you have to try to breed your own. Because of the very low ratio of show stock bred, they are best kept by fanciers who have the facilities for keeping them in very large numbers, although some fanciers keep a few as a second string.

ESSENTIAL REQUIREMENTS

So, what do you have to breed, to earn the acclaim that all top tort and whites can expect? **In short you want a smooth, short-coated cavy, with patchwork of red, black, and white. The patches should be as square as possible with straight demarcation lines between the patches.** The opposing patches on each side of the body

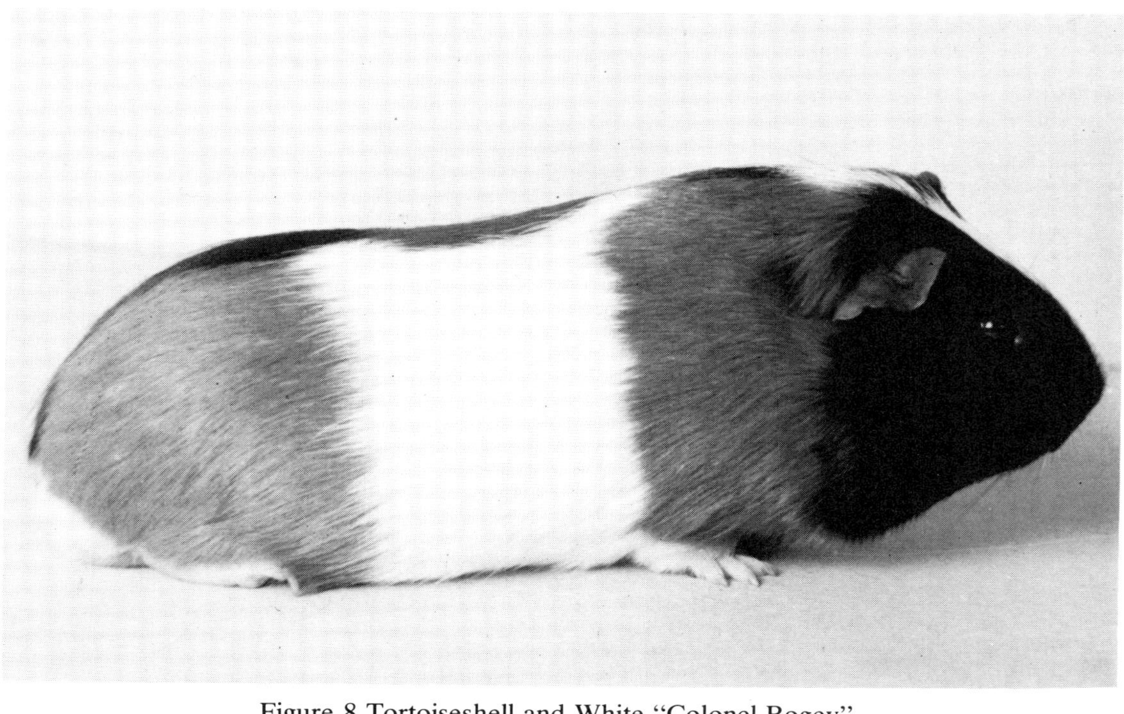

Figure 8 Tortoiseshell and White "Colonel Bogey"
(Owned by Mr. George Howard)

should be of different colours. The demarcation line between the sides should run in a straight line, from the tip of the nose, between the ears, down the centre of the back to the rump, between the legs, down the centre of the belly and under the chin to the nose again. Each side of the head should be of a different colour—ideally without a blaze, but if a blaze is present, then there should be different colours on either side of the blaze.

FAULTS TO AVOID

It used to be said, when talking of faults in a Tort and White, that you must avoid "the three 'B's"—Breeching, Banding and Brindling. *Breeching* is when one colour runs right around the rump—like a pair of breeches! *Banding* and *Belting* means a band of one colour going part or the whole way round the body. *Brindling* is patches of hairs of intermingled colours.

It does not matter how many patches the Tort and White has so long as it has all three colours on either side. However, to have the patches as square as possible, it is best to have four patches on either side, including the head.

SELECTING BREEDING STOCK

Then there arises the problem of selecting breeding stock, and this can be difficult to resolve. Like the Dutch, the Tort and White is basically a white cavy with coloured markings, so theoretically, the white will increase with each generation; therefore, it is best to choose animals with a small amount of white. However, one well-known breeder told me that he obtained all his

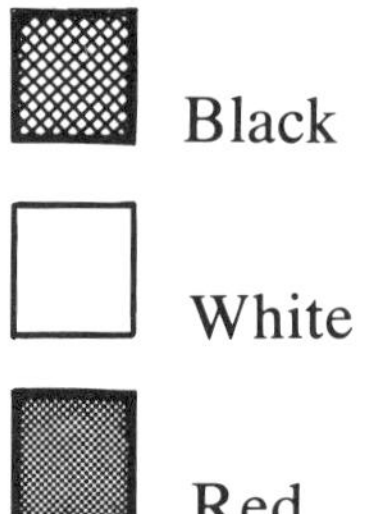 Black

White

Red

1. This cavy is banded white—that means a band of all the way round the body.
2. It is breached black—the black goes all the way round the rear.
3. Red patch is much too long.
4. White patch is quite good.
5. Almost solid head.
6. Overlap.
7. Uneven dividing line underneath.
8. Left side lacks red.
9. Right side not enough white or black.

See Diagrams opposite

FAULTS IN TORTOISESHELL AND WHITE
See opposite for key.

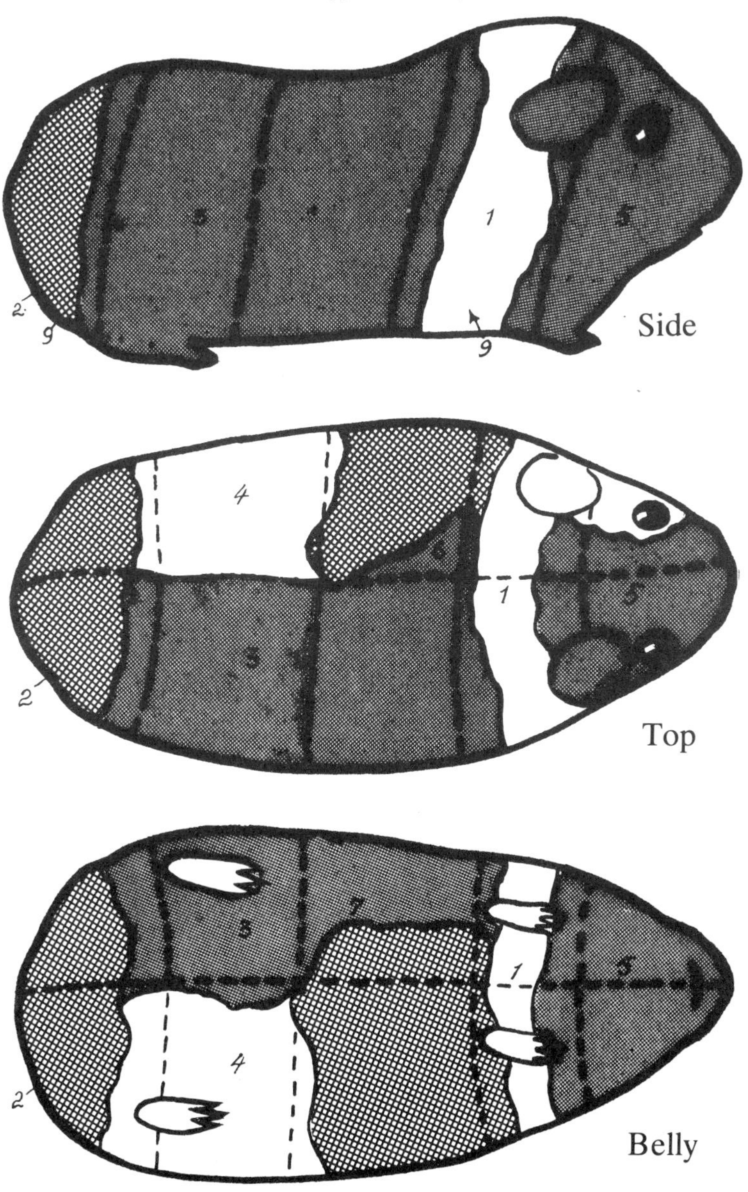

best-marked animals from a sow which was all-white on the body with just a little colour on the head. It is, however, best to avoid certain faults in breeding stock. At the top of the list, I would place brindled patches, followed by white banding, and those with Dutch marked heads. Richness of colour is another point to watch. No matter how well-marked a cavy may be, if the colour is too pale and washed-out looking, it will lose points.

Look for breeding stock with clean cut patches, even if the cut is in the wrong place. Look also for a nice straight line down the belly, even if it should be uneven on the top. If a sow lacks one colour on one side, mate her with a boar in which the missing colour predominates *on the same side*.

To breed a perfect Tort and White is every breeder's dream, but it has never yet been achieved; even the best can be faulted. So, do not hesitate to show "near-misses" and, maybe, one day you could be the one to achieve the dream!

Standard for Tortoiseshell and White Cavy

Patches—to be clean cut, clear
and distinct 25
Patches—equal distribution and
uniform placing thereof 25
Colour—black, red, and white..........20
Shape and size 15
Eyes and ears 5
Coat and condition....................10
                                     ———
Total 100

Faults

Side whiskers, bands, belts—Cavies being short of any coloured patches on either side shall be penalized by twenty-five points.

CHAPTER 10

The Himalayan Cavy

Very often the Himalayan is likened to the Siamese cat which it resembles, with a light-coloured body and dark points.

In the case of the Himalayan cavy, the body should be as white as possible, and the "points", which are the nose, ears and feet, should be as dark as possible, in either of the two recognised colours—black and chocolate. Even these two colours may be confused, and on occasions, a poor-coloured black may be mistaken for a chocolate. The black is *not* a true-black, but may look like a dark *plain* chocolate colour, especially at the edges of the coloured points, and a black whose points have faded can be deceptively like a chocolate. In this case the colour is peppery, or flecked throughout. In the case of a true chocolate, the colour of the points is milk chocolate or dark beige, and should have no flecking. If in doubt look at the colour of the ears, because they, more than anything, give the true colour and do not fade at intervals like the nose and feet. The eyes are red.

BREEDING THE POINTS
You will notice my use of the phrase "fading off". I am

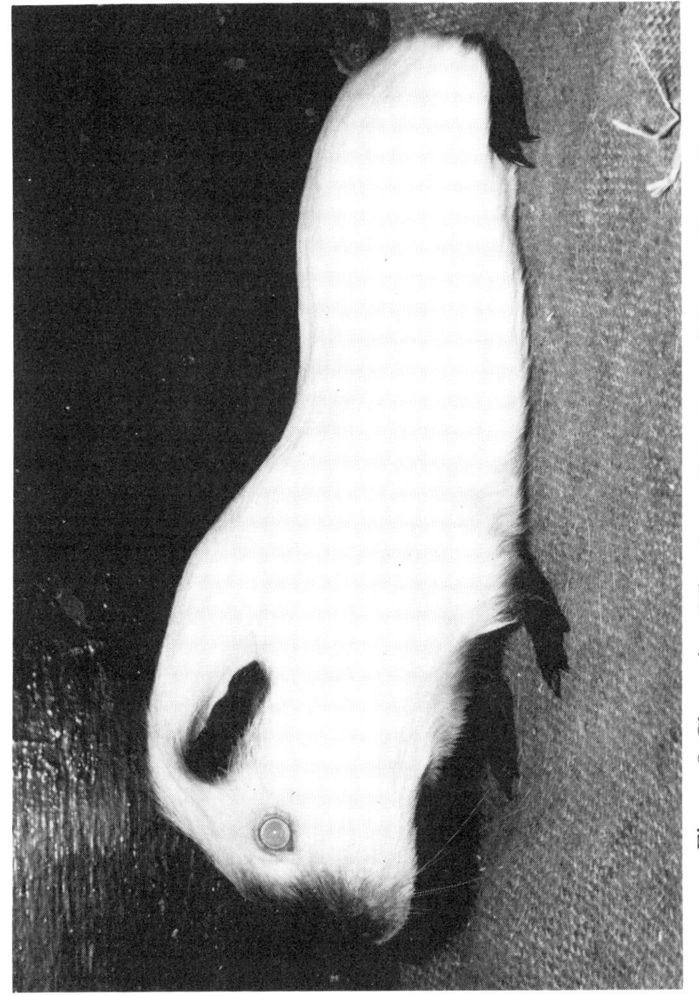

Figure 9 Champion Himalayan "Black and White Minstrel"
(Owned by Mr. A. E. Roebuck)

afraid that this can happen quite frequently to the points of Himalayans, and is mainly due to outside influences. I refer, of course, to *good* Himalayans and not to the poor quality animals which may never darken properly on the points. The colour is always at its best when the cavies are kept in cool, dry, shady conditions. A hot summer can mean that there are no dense pointed Himalayans around at all. Then, as the weather becomes cooler, they emerge in all their glory. Bright sunlight, damp, and injury, etc., can also make the points fade, but extreme cold can make the body colour much too dark, so the conditions in which Himalayans are kept is extremely important. I keep my Himalayans in the lowest tier of hutches, facing away from the light, and if the shed becomes very hot during the summer, I move them out to an open—yet shady—brick out-house which remains cool.

Himalayans are pure white when born, and this sometimes makes beginners think they have been "done" when their first litter arrives, but within a day or two you will notice the pads of the feet darkening, and gradually the points will come "through" and the white hairs will gradually disappear from the feet, the nose will darken at the tip, and gradually, the dark patch will spread up the nose to between the eyes. At about five to six months of age, the points will be clear.

FAULTS

Faults to look for in youngsters at this stage are white toes or toe nails, white patches on the feet and flesh coloured patches on ears. The nose smut should go well

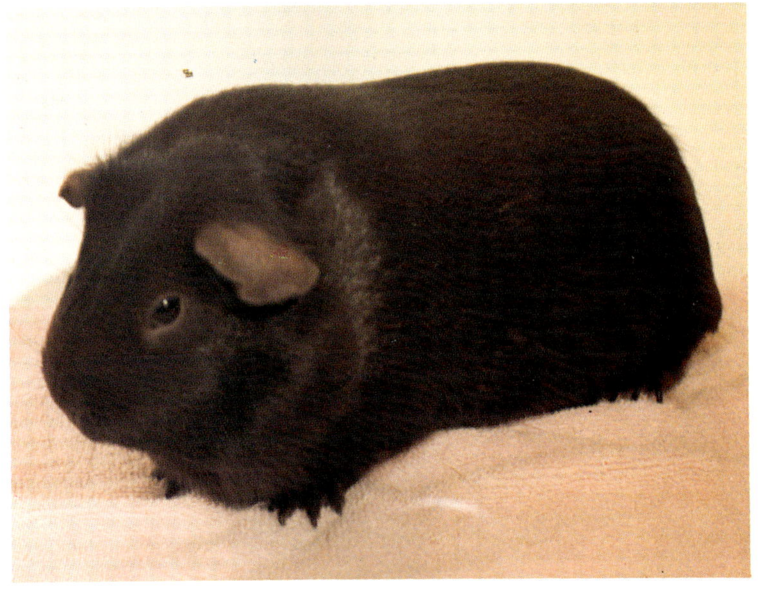

Plate 2   *Top*   White Self Cavy (Mrs M. Pearce).
                    Bottom   Chocolate Self (Mrs M. Pearce).

up between the eyes. At this age, young Himalayans usually have good body colour—usually pure white. If they tend to be dark on the rump, then the chances are that they will be too dark as adults. But do not discard these dark-backed ones—especially sows—as they are useful breeders. The ones to discard are the ones with white patches on feet and ears.

CORRECT SIZE

Type and size in Himalayans have been controversial subjects in the past. Some breeders try to breed them larger and with the type of a self, but although occasionally a dense-pointed one comes along among these, they usually fail badly on points.

To me, the Himalayan is a small, neat cavy, long in body compared to other varieties, and while I do not like sharp-pointed noses, neither do I like the tennis-ball shaped head. The chocolates tend to be larger than the blacks, but the Himalayan, with its pure white body and sharply contrasting points, is attractive and dainty, and is greatly admired at shows.

Standard of the Himalayans
Black Himalayan
    Nose—even and jet black 25
    Feet—jet black 20
    Ears—shapely and jet black 10
    Coat and colour—short, silky
        and pure white 20
    Size and shape 10
    Eyes—large, bold, ruby red 5
    Condition 10

                        ———
    Total 100

Chocolate Himalayan
    Nose—even and rich milk chocolate 25
    Feet—rich milk chocolate 20
    Ears—shapely and rich milk chocolate 10
    Other points as for Black Himalayan

CHAPTER 11

The Peruvian Cavy

The Peruvian is the most exotic of all the varieties of cavies. It is the one which attracts the eye of the public at the big shows, and many of them just do not believe it is a cavy at all.

Genetically, the Peruvian is a rosetted, long-haired cavy. The rosetting makes the hair on the top of the body lie towards the head and fall over the face, while the hair on the rear end falls over the hindquarters making it difficult to tell which end is which.

NOT FOR THE NOVICE

This is not a beginner's cavy, as it requires a great deal of careful grooming to get it in "show coat" and then to keep it that way. Although it is a rosetted cavy, it is best for it to have only the two rump rosettes, which control the lie of the hair. An odd rosette on the side may be covered as the hair grows longer, but a rosette on the top of the back may spoil the cavy for show purposes.

So, when you have a baby Peruvian, which you hope to groom for showing, it will look like an animated powder-puff. The first thing you must do is train it to sit still. When it is very small, let it sit on your hand, while

Figure 10 Peruvian Cavy
(Owned by Mrs. Pat Wood)

you stroke it *from the rump to the head*. It will soon learn that you mean it no harm, and will enjoy the attention. As it grows bigger, sit it on a box or stand covered with cloth—a shoe-box is ideal—and brush it with a soft baby brush.

HAIR CONDITIONING

The hair of a Peruvian grows at an average rate of about an inch per month, and if you want to get him in show coat, it is necessary to protect the coat from wear and tear and keep it clean. When your little cavy is about four to five months old, the coat will be trailing a little at the back end, or "sweep", as it is called; so it is now time to start wrapping it (see diagram). At this stage, a wrapper can be put in the sweep. Each day the coat must be brushed out and the wrapper replaced. You may find, that at first, the wrapper will not stay in, but you must be patient and replace it.

About a month later, you will have to put a wrapper in each side, parting the hair down the centre of the back. Now daily grooming must continue and the wrappers replaced, and you will see why it is important to train them to sit still. If a cavy objects very strongly to having wrappers, and will not sit still, you might as well give up that particular cavy. The wrappers are light, and cause no discomfort if they are put in properly, but you do, on occasions, get a cavy that is temperamentally unsuited for showing.

As the hair grows longer you may have to enlarge the size of the wrappers.

It is obvious that wrapping is only for show stock.

continued on page 81

TO WRAP A PERUVIAN OR A SHELTIE

Stage 1

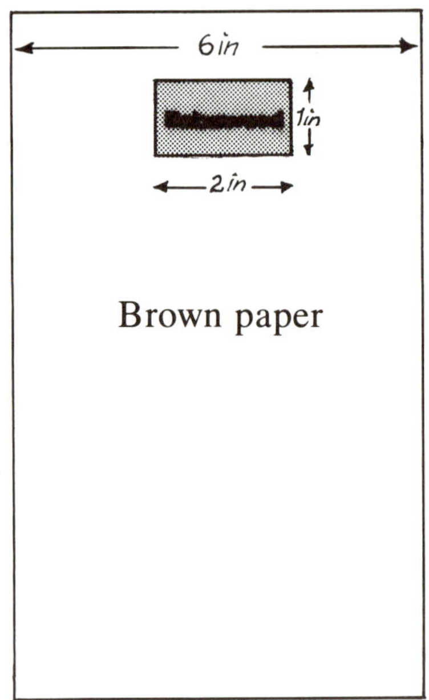

Stage 2

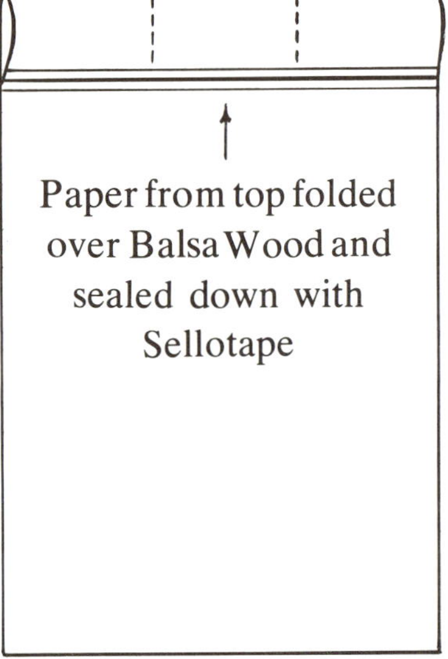

Stage 3

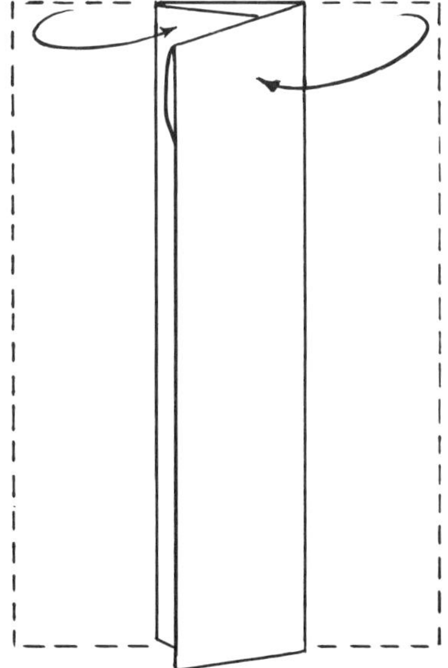

Fold sides over centre panel

Stage 4

Fold concertina-wise

Stage 5

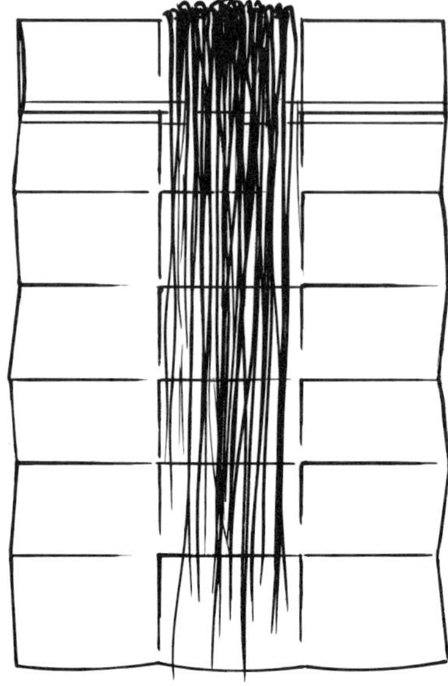

Unfold paper and place under the cavy's hair.

Stage 6   Fold up again and secure by an elastic
           band.

One wrapper on the sweep or tail-end, and one
on each side is sufficient for a time.

You will find that you will have to enlarge them
and increase the number as the cavy's coat grows.

Breeding stock should have their coats clipped, but they do need to be brushed and bathed, because the coats can quickly become matted, even when clipped.

AVOIDING HAIR DAMAGE

A bad habit which some Peruvians have is that of chewing their own and their hutch-mates' coats, and there is nothing more heart-breaking than to spend months grooming only to find one morning that your pride and joy has given himself a "short back and sides". To help to prevent this, here are a few tips, although there are a number of theories on why they chew their coats.

1. Be sure that wrappers are in comfortably.
2. Keep show cavies singly in hutches.
3. Be sure they have plenty of hay to chew—either chopped up in short lengths or tied in a bunch on the netting wire to prevent it becoming tangled in the coat.
4. If possible keep your show Peruvian in the same hutch all the time. If you change hutches, be sure they are thoroughly cleaned and disinfected, so that your cavy does not get the smell of a strange cavy which may cause him to chew. Also have a separate brush and show stand for each cavy.

SHOW STANDS

If you are showing Peruvians, you should supply yourself with show-stands, which are covered little platforms about eighteen inches square, about six inches high, and covered with hessian—easily made at home. The

Peruvian stands on the show stand, wrappers are removed, and its coat is brushed out. It is then carried on to the judge's table. This prevents its coat being tangled when it is carried. When you see an adult Peruvian, whose coat may be over twenty inches long and very soft and silky, you will realise why all this is necessary. There are no points for colour in the standard in Britain; the emphasis is on coat quality and they may be any colour—or combination of colours.

Show Standard for the Peruvian Cavy

Head broad with prominent eyes 5
Fringe, with hair completely
    covering the face 15
Shoulders and sides 15
Texture to be of a silky nature 20
Density 15
Sweep, length and fullness, the hair
    falling over the hindquarters 15
Condition 5
Size 5
Presentation 5
                             ———
Total 100

Notes

The hair should be fine, silky and glossy.

The fringe should be furnished so that the hair hangs in a thick mane on each side of the head.

The face should be short, the eyes full and large.

While we aim for a straight coat, a slight wave should

not be unduly penalised.

If the sweep is slightly longer than the sides, this does not constitute uneven length.

A SHOW STAND

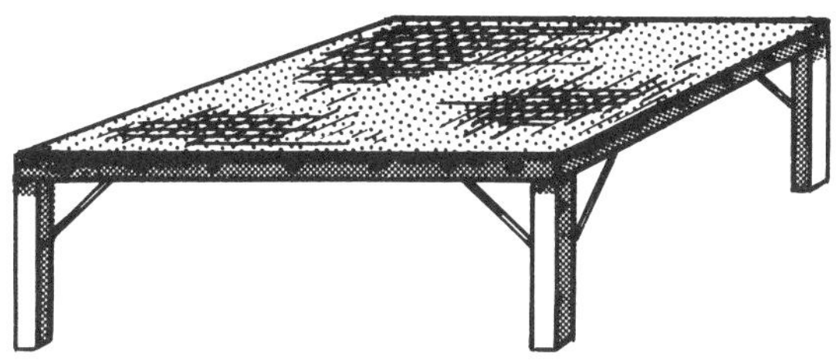

A simple show stand for Peruvians
and Shelties covered with hessian.
16 inches square and 6 inches high.

CHAPTER 12

The Sheltie Cavy

It is only a very few years since the Sheltie cavy was taken up seriously as a show cavy, but you only have to see some of the beautiful Shelties there are today, to realise the potential of this lovely variety. It is known as a "Silkie" in America, where it is not as yet accepted as a show variety, but some enthusiastic American breeders are trying hard to have it accepted there.

A SMOOTH LONG-HAIRED VARIETY

Like the Peruvian, it is a long-haired cavy, and needs as much care and attention in grooming and wrapping. The difference between them is that the Peruvian is the long-haired "rosetted" cavy, and the Sheltie is the long-haired "smooth" cavy. Not having the rosettes to make the coat lie towards the head, the Sheltie's hair flows backward leaving the head clear, and forming a long "train" at the rear. Many people say they prefer it to the Peruvian, because its face is visible, and not hidden by hair.

Nowadays the Sheltie has as much length and density as the Peruvian. Instead of a central parting, taking the hair to each side—as with the Peruvian—it should be brushed straight back to join the sweep.

BREEDING

To understand the Sheltie, you need to know a little about genetics, with special emphasis on Mendel's pea experiment and dominant and recessive genes. To put it simply—when a Peruvian, with its dominant rosetting genes and recessive long hair genes is mated to a smooth *(recessive)* short coated *(dominant)* cavy, the resulting litter are what we call "fluffies". But, interbreed with these, and the genes will sort themselves out in various ways, but an average of one in sixteen will inherit the two recessives—smooth coat and long hair. This is your first generation Sheltie. As these are recessive for both properties of coat, Sheltie bred to sheltie will breed only Shelties. However, there are enough good Shelties around now for you to buy animals which have been Sheltie bred for several generations.

YOUNG SHELTIES

Baby shelties look quite different from the powder-puff baby Peruvians. They look just like baby smooth, short-coated varieties, but then very soon you have the pleasure of watching their coats grow.

It is necessary to train them in the same way as described for Peruvians, and they also require to have their coats wrapped if you want to show them (see page 78). At first many people only put a back wrapper in Shelties, but nowadays side wrappers are put in as well. When you brush them out, then brush all the hair backwards to blend with the sweep.

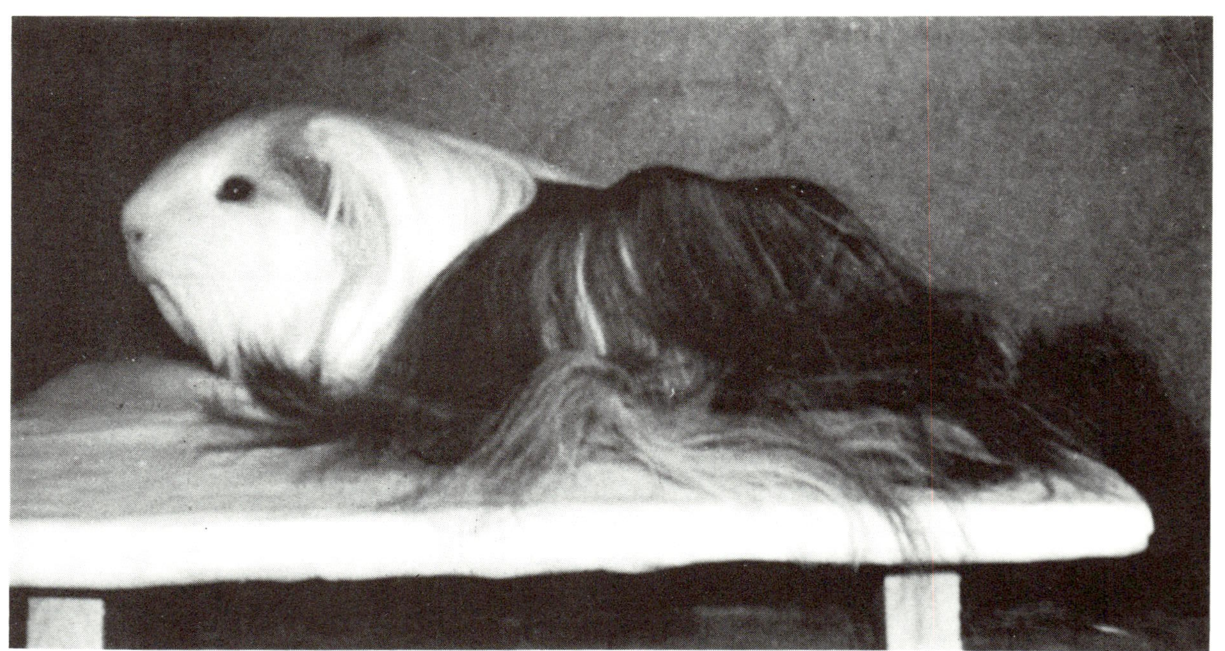

Figure 11 Champion Sheltie "Highwayside Humphrey" (Side View)
(Bred by Mr. and Mrs. G. Colley)

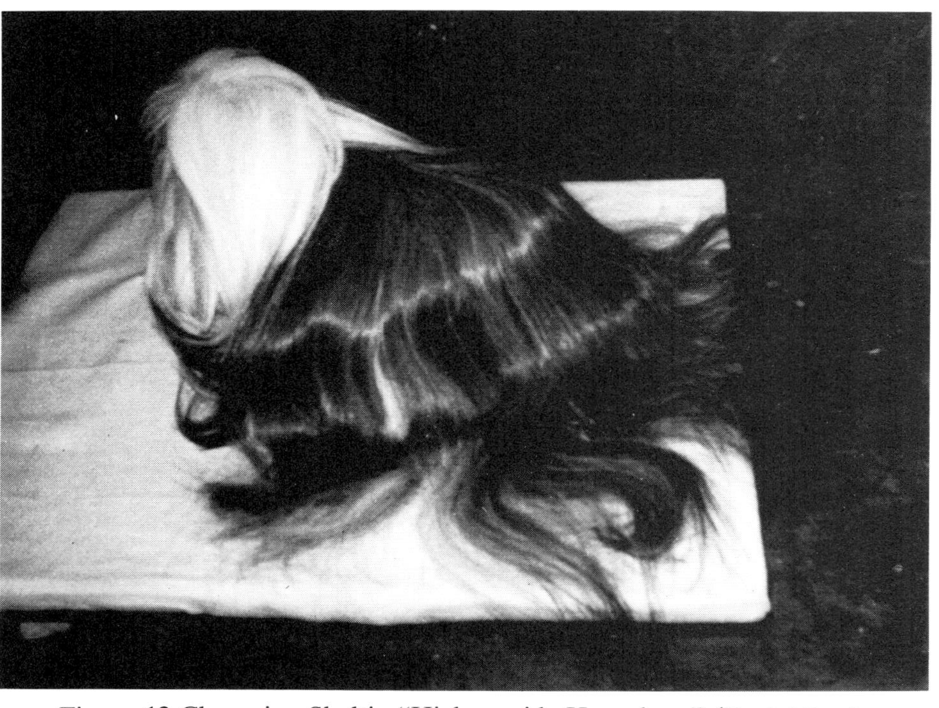

Figure 12 Champion Sheltie "Highwayside Humphrey" (Back View)
(Bred by Mr. and Mrs. G. Colley)

The Sheltie also requires show stands as described for Peruvians, and, of course, they too have their coats clipped for breeding. Again, like the Peruvian, the Sheltie may be any colour or combination of colours.

Standard for Shelties

<pre>
Head, broad with short nose and large
    prominent eyes with hair lying towards
    rump. Ears to be petal shaped, set
    slightly drooping; good width between . 15
Mane—sweeping back to join with sweep;
    is not parted 15
Shoulders—broad with hair of good length
    and density, continuing evenly around
    sides 20
Coat—silky texture and good density 20
Sweep—length and fullness of hair falling
    over hind–quarters; to be longer than
    sides which should be even 20
Condition and presentation; with no
    parting 10
                                          ———
Total 100
</pre>

CHAPTER 13

The Crested Cavies

I sometimes feel a sense of wonder, when I reflect that I was responsible for introducing the crested cavies into Great Britain, when I imported six of them from Canada in 1972. They have become very popular and it looks as if, in the future, they will outstrip all the other varieties.

CREATIONS OF THE U.S.A.

Cresteds were first bred in the United States of America, where they appeared as mutations in litters of smooth-coated cavies. Instead of discarding them, the fanciers bred with, and developed them. When bred with either other cresteds or smooth-coated cavies, they produce about fifty per cent of crested off-spring.

I have no doubt that this mutation has cropped up on other occasions, and will do so again. I, myself, have bred, quite by chance, a crested youngster, from a pair which I knew quite definitely had no crested blood. If I had not been interested in cresteds, the probability is that I would have discarded it as faulty. So we must give the American breeders the credit for first developing them as show cavies.

The crest is in the form of a rosette, situated centrally

Figure 13 American Crested Champion "Narinia Tonto" (Front View)
(Owned by Mrs. Isabel Turner)

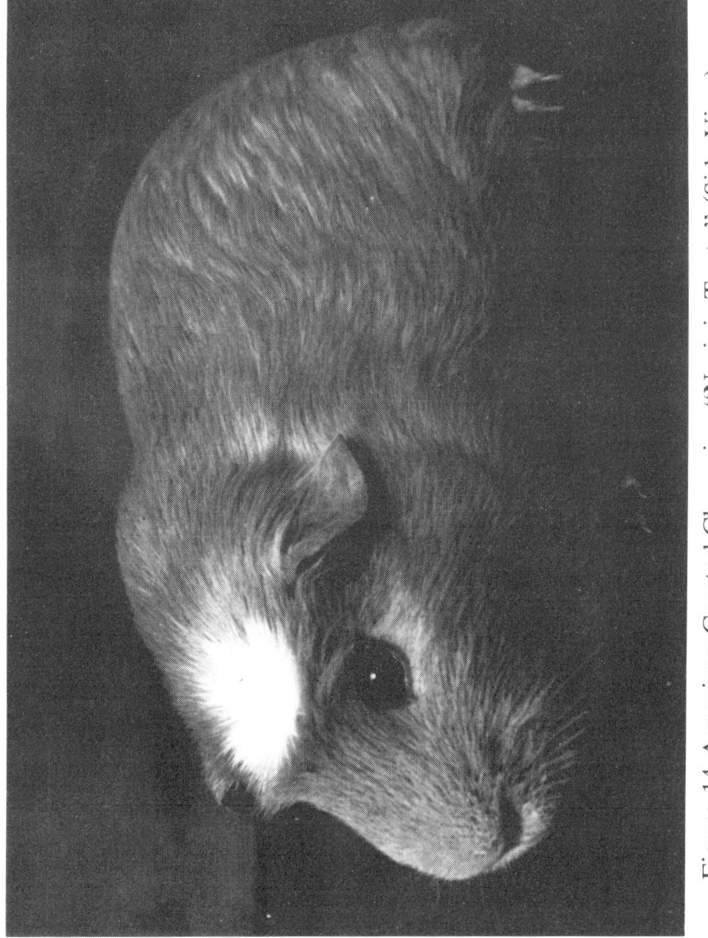

Figure 14 American Crested Champion "Narinia Tonto" (Side View)
(Owned by Mrs. Isabel Turner)

above the eyes and below the ears. The hairs of the crest radiate from a central point and give the cavy the appearance of heavy eyebrows and a frowning expression. Apart from the crest, the cavy should be smooth-coated.

AVAILABLE COLOURS

The American Crested, so called because it is the type developed and recognised as a show cavy in the States, should have a self-coloured body and a crest of a contrasting colour. The most common colour is white-crested golden. The originals of these in this country were dark-eyed, but many breeders are concentrating on producing the red-eyed golden with a white crest.

There are some nice white-crested creams, but care should be taken to be sure that the crest is really white, and not just showing the paler undercolour of the cream.

Many other colours are being produced, but good show specimens are few and far between. White-crested blacks are proving difficult to breed, although there are quite a number of near misses. One of the difficulties which crop up with these is brindling or red hairs intermingled with the black, but considering the few years that they have been bred, I am sure that with a little more time, these difficulties will be overcome.

It is possible that we shall see American Cresteds with more variety in the crest colour—such as red-crested blacks and cream-crested chocolates.

ENGLISH CRESTED

English Cresteds are self-coloured cavies with the crest to match the body colour. They are being bred in all the colours of the Self variety. It is among these that there have been some really outstanding specimens, especially among the blacks.

HIMALAYAN CRESTED

Himalayan Cresteds, are, as you would imagine, Himalayan marked cavies with a crest. At the present time, they do not have enough density of colour on the points, but they will improve.

AGOUTI CRESTED

Agouti-cresteds is another variety which needs quite a lot of selective breeding to improve it. It is mainly found in lemon and chocolate agouti colours.

These are the varieties most commonly seen, but the Crested Cavy Club and the British Cavy Council have now ruled that any variety which is standardised in its smooth form is accepted as standardised in its crested form.

The Guide Standard Cresteds, for example Argente, Harlequin, Saffron and Dark–eyed Golden may be shown in Guide Standard or Unstandardised classes and from January 1980 will be catered for by the Rare Varieties Cavy Club. The Crested Cavy Club will also cater for them at their Stock Shows, until such time as the smooth varieties become standardised, when the crested varieties will automatically follow.

The Crested Sheltie is dealt with elsewhere in this book.

The varieties which already have had rosettes in their genetic make-up (which incidentally, is an entirely different gene from the one which gives the crest)—the Abyssinian and the Peruvian, are not at all likely to produce successfully a crested variety, and again, what would be the point of it?

THE STANDARDS
Standard for the English Crested Cavy

Crest, to match body colour	20
Colour, to conform to the colour of the matching English Self	24
Shape, short cobby body, deep broad shoulders	20
Coat, short and silky	12
Ears, Rose-petal shaped, set slightly drooping with good width between	8
Eyes, large and bold	8
Condition	8
Total	100

The crest to radiate from a centre point between the eyes and ears.

The crest to be a deep rosette, the lower edge to be well down the nose.

Any different coloured hairs in the crest to be severely penalised.

Abundance of different coloured hairs on body to be penalised as in self cavies.

Owing to the difficulty of getting the correct markings of the crest of the American, the points of crest have been increased.

Standard for American Crested Cavies

Crest, to be a contrasting colour
    to the body colour 30
Colour, body colour to conform
    to matching English Self 21
Shape, short cobby body,
    deep broad shoulders 18
Coat, short and silky 10
Ears, rose-petal shaped, slightly
    drooping with good width between 7
Eyes, large and bold (in the golden may
    either be red or dark) 7
Condition 7
                                                           ———

Total 100
The crest colour to be as near to a complete circle of
solid colour as possible. A circle of less than seventy-five per cent to be severely penalised. The crest
colour should not appear elsewhere on the body. A
blaze of the crest colour to be severely penalised.
Hair of the body colour in the crest to be penalised.

Himalayan Crested

Crest, to be white to match
    body colour 20
Colour, to conform to colour
    of matching smooth Himalayan 24
Other points as for English Crested

Agouti Crested

    Crest, to match body colour.............20
    Colour, to match colour and pattern
       of corresponding smooth agouti 24
    Other points as for English Crested.

The Crested Sheltie

I have always maintained that it is a mistake to breed crests into a variety such as the Sheltie, whose beautiful coat takes attention away from the crest. However, the crested Sheltie is now here to stay, and has a large and enthusiastic following. These cavies are certainly very attractive, and their breeders have decided to form their own specialist club. Details of this club are not yet available. The name of the cavy may be changed to avoid confusion with both the Crested Club and the Sheltie Club. Because of the formation of this new club, Crested Shelties should not now be shown in Crested Classes. If there is no class specifically for them they should be shown in the A.O.V. class.

The long hair gives this variety a very large crest, and combined with a short broad face, large eyes and long and dense hair, it presents a very attractive cavy, and makes a third long haired variety. Grooming and wrapping is the same as for the Sheltie cavy.

Standard for Crested Sheltie

    Crest, radiating from a centre point between
       the ears and eyes 20

Figure 15 Crested Sheltie 'Lizzy'
(Owned by Rachel Gibbons)

97

Head—broad, short nose, large prominent
  eyes and hair lying towards the rump .. 10
Mane—sweeping back to join the sweep;
  not parted 12
Shoulders—broad; hair slightly longer
  continuing along the sides at equal
  length 16
Sweep—length and fullness of hair falling
  over hindquarters 16
Coat—silky texture; good wealth of coat
  at sides 16
Size, condition and presentation 10
                                        ———
Total 100

CHAPTER 14

Roans and Dalmations

Please note the spelling of Dalmation; this is the spelling requested by the Roan and Dalmation Club and is not a mistake.

Although these two varieties are classed together and have a joint specialist club, they are not related.

The smooth roan, the original being the blue roan, was developed after years of selective breeding by Jon Belling of Fillongly near Coventry. The resulting cavy is an attractive mixture of black and white hairs on the body, with black head and feet.

The Roan and Dalmation Club proposed, and it was accepted by the British Cavy Council that all standardised self colours be accepted as standardised in Roan and Dalmation and that they be renamed to correspond with the self colours. Thus our old blue roan is now called 'Black Roan' and so on. The black roan is still the one most often seen at shows but there are a few red and golden roans (the former strawberry roans) to be seen.

Some of the faults encountered in roans are as follows: uneven roaning, dark patches on rump. light bellies and odd coloured feet. There are, however, some very nice ones being shown, and on occasions these have even reached Best-in-Show.

The Dalmation: this is a spotted variety, and is accepted in any self or agouti colour. It evolved from a mutation from self blacks and was originally bred by Elizabeth Wilson of Harrogate. They should have black or other corresponding colours, or silvered head with a blaze, ruby eyes, and coloured or silver feet, to match the head.

Faults include: lack of a blaze, spotting too heavy, uneven or too light. Dappled roans are now considered to be mismarked (too heavy); Dalmations are not accepted as show cavies.

Spots on the belly are an advantage, but some of those that are beautifully spotted on the top, lack belly spots, but should not be too severely penalised for this.

Some difficulties arise if Roan are bred to Roan or Dalmation to Dalmation. This will result in about one in four of the babies being born with small, practically non-existent eyes. These are pure white and are called 'Microthalmic Whites'. They usually do not live long and are best culled at birth. To avoid breeding these microthalmic whites, breed Roans and Dalmations to the self cavies which are bred from these two varieties.

Standards

Black Roans

    Roan, mixing to be even throughout 30
    Head—clean cut and solid black 10
    Feet—solid black...................... 10
    Eyes—black, large and bold 10
    Coat—to be short and silky; colour black
        and white 10
    Ears—black, set wide apart, large and
        drooping 10
    Size, type and condition................. 20
    Total 100

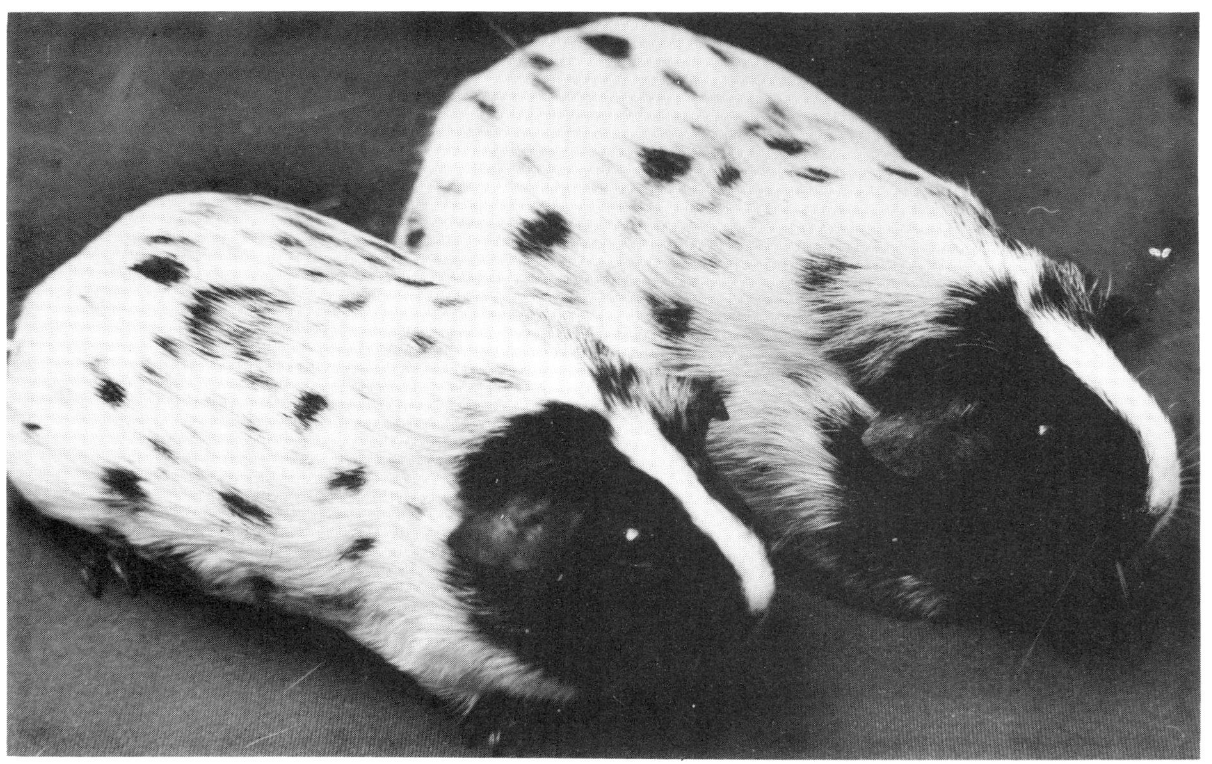

Figure 16 Dalmations *(Bred and owned by Janice Lawson-Reay)*

Remarks

The Roan is basically a black cavy with white hairs evenly intermixed throughout the body. Solid black should be confined to the head and feet, but white whiskers should not be penalised.

The type and overall shape of the Roan should ideally conform to the requirements of the Self Cavy. Eye colour is black but, in certain lights this may appear to be ruby. This is not a fault.

Coloured Roans

All coloured Roans must conform to the respective coat colour requirements set down by the E.S.C.C. or in the case of Agoutis the N.A.C.C.

Disqualifications

Fatty eye, side whiskers.

Dalmations

Spotting—black spotting on white body,
clear and distinct, well distributed over
body 30
Head—white blaze, black either side,
broad, with roman nose 20
Legs and feet—black or slightly silvered
on all four legs...................... 10
Eyes—large and bold, colour deep ruby .. 10
Ears—black, set wide apart, large and
drooping 10
Shape, coat, condition and colour: to conform to the requirements of the

self cavy 20
Other colours: to conform to the require-
    ments of the matching Selfs or Agoutis

Total 100

CHAPTER 15

The Rare and New Varieties

A few years ago, there were only two varieties which came into the category "Rare Varieties". These were "Tortoiseshells and Brindles".

"The Rare Varieties Cavy Club" was formed to cater for these two varieties, which were accepted as standardised in more recent years. The R.V.C.C. has now taken under its wing any new variety which is considered worthwhile.

The Rare Varieties can be sub–divided into:
1. **Standardised**
2. **Guide Standard**
3. **Unstandardised**

STANDARDISED VARIETIES

The Standardised Varieties are in general those that have reached a standard of excellence, which the R.V.C.C. and the B.C.C. (British Cavy Council) have accepted and provided a standard complete with points for each feature.

The Tortoiseshells, and the Brindles were standardised long before the R.V.C.C. and the B.C.C. came into being, and are so few in number that they look like being rare varieties for evermore.

When breeders of a standardised rare variety feel that they are strong enough in numbers, they form a specialist club for the variety. Before they can do so however, they

Plate 3   *Top*   Cream Self Cavy (Mrs M. Pearce).
                      Bottom   Tort and White (Mr J. Tenner).

must apply first of all to the club which caters for them, usually the R.V.C.C. but in the case of the Crested Sheltie it was the Crested Cavy Club, and then take their proposal, through the representative of that club to the B.C.C. If they accept and pass the proposal then they can go ahead and form their new club.

A variety can, in the same way be taken over by an existing club. The dark–eyed white was adopted by the English Self Cavy Clubs, but at present they do not accept the D.E. Golden, the Saffron or the Buff. These should thus be shown in Rare Variety classes. However, nowadays under British Cavy Council rules a variety must first go through the Guide Standard Stage before being given a full standard. More of that later.

The Standardised Rare Varieties

Tortoiseshells and Brindles

These are the two oldest varieties which are still classified as Rare Varieties, and they still appear to be making very little progress. Now and then a few good ones appear, but they are very sparse in number, and are only being bred by a handful of dedicated breeders.

Tortoiseshells

A Tortoiseshell—usually called a "Tort" should have clearly defined patches of red and black, similar to the Tort and White, but minus the white. Torts are even more difficult to breed to show standard than the Tort and Whites. The genes which cause the white markings on cavies appear also to have the effect of breaking up the colours into well defined patches. Take the white patching genes away, and the patches become very indistinct with the red and black hairs intermingling. So it would seem that the only way to breed a well-patched Tort, is to have a patch of white somewhere on the cavy. But then it could be classified as a "mismarked Tort and White"; so, what to aim for, is a cavy with black and red patchwork, with a white toe or a tiny white patch, say just a few hairs, on the belly. A white patch on top would stand out too clearly. No wonder it is difficult to breed!

Brindles

These should have an even intermingling of black and red hairs all over. It is difficult to achieve even brindling, so you find light and dark patches. The breeders of these two varieties—and now and then they produce a good one—deserve a great deal of admiration for their perseverance.

Tricolours and Bicolours

Tricolours are patched as a Tortoiseshell and White but in different colours. Some attractive examples are chocolate/cream/white, and black/cream/white, etc.

Bicolours are patched as a "Tort" but in different colours which include black/white, chocolate/cream, etc. Neither of these are in sufficient numbers as yet to have any likelihood of leaving the Rare Varieties for some time.

THE STANDARDS

Tortoiseshell

Patches, clear and distinct 45
Eyes, large and bold 10
Coat, short and silky 10
Size, shape and condition 20
Colour, black and red 15

Total 100

The colour should be black and red, equally distributed in distinct patches, the smaller and more uniform the better.

Brindles

Evenness of brindling 45
Eyes, large and bold 10
Coat, short and silky 10
Size, shape and condition 20
Colour, red and black 15

Total 100

Brindling of coat to be even, and free from patches of distinct colour.

Tricolour

Patches, clear and distinct 45
Eyes, large and bold 10
Coat 10
Size, shape and condition 20

Coat, any three colours
    (other than tort and white) 15
                                           ———

Total 100

Bicolour

Patches, clear and distinct 45
Eyes, large and bold 10
Coat, short and silky 10
Size, shape and condition 20
Colour, any two colours
    (other than red and black) 15
                                           ———

Total 100

GUIDE STANDARD CAVIES

The British Cavy Council would like to discontinue "Unstandardised" classes at shows, in favour of guide standard classes. In this way judges would at least have a guide to what to look for in a new variety. However, many clubs still put on unstandardised classes; some without the Guide Standard Class, and some in addition to it. If there is no guide standard class then it is quite in order to show guide standard cavies in the unstandardised class, as they are not considered as fully standardised. Before a guide standard is accepted it must be put forward by the R.V.C.C. to the British Cavy Council. They can and very often do, turn it down.

These are the standards that are already accepted:

Argente

This is really a red–eyed agouti. The colour accepted by the B.C.C. is lilac based with golden ticking. There are other colours but they are not yet accepted. Application was made to the B.C.C. to have the beige based Argente accepted but this was turned down, but no doubt application for that and other colours will be made again.

At the 1979 A.G.M. of the R.V.C.C. it was proposed to discuss the change of name, as 'Argente' really means silvered, and these varies are *not* silvered. However, at present nothing further has been done.

There are some silvered cavies, which will be discussed in the unstandardised section.

Harlequin

This is a patchwork cavy. The patches are of two colours and a roaned or brindled mixture of the two. For example; black, white and blue roan (known as the Magpie) or chocolate cream and choc/cream roan.

The head markings should be half of the head one colour and the other half the other colour or roan. With a straight dividing line down the centre of the face. Feet should be alternate colours; front feet one of each colour, back feet the same but at opposite sides. This is a very difficult cavy to breed, but a challenge.

Dark–eyed Golden

The Dark–eyed Golden is a self, but has never been recognised by the English Self Cavy Club. They should be as for the Red–eyed Goldens but with dark eyes which should be black or ruby. One of the commonest faults is pigmentation of the skin on ears and feet, but they very often excel in type—more so than their red–eyed counterpart.

Saffron

This is also a self which has not been accepted by the English Self Cavy Club, but it is still fairly new. It could be described as a red–eyed Buff, but should be rather brighter in colour than the Buff and more lemon coloured. At present, they are still not up to some of the Selfs in type but they are improving all the time.

GUIDE STANDARDS

No points are allocated in guide standards. They just give judges and breeders an idea of what to look for.

Argente

Type—of good type; a cobby shape with deep shoulders, broad head and short nose.

Ears—large and drooping

Eyes—large, bold and pink in colour

Colour—undercolour lilac; the hairs tipped with gold giving the coat a look of shot silk; belly golden; skin colour matches the undercolour. Overall appearance a very attractive type of cavy with a smooth silky coat that has an even sheen

Harlequin (Black Harlequin)

Head—half predominantly black, half yellow, divided down centre of face

Body—3 colours each side, with equal balance

Patches—of black, yellow and black/yellow brindle; straight line top and under; square patches of uniform size

Serious fault—completely solid, one side

Faults—belted, absence of one colour on one side, white toes and white hairs (except in Magpies)

Disqualification—white on body, white leg or large white patch on head (except in Magpie)

Colour—black, yellow and black/yellow brindle

Type and size—general cavy type; large as possible

Eyes—dark large and round

Ears—large, well-set and drooping

Coat and condition—

Other colours

Chocolate Harlequin: Chocolate/yellow and choc/yellow brindle

Black Magpie: Black/white and black/white roan

Chocolate Magpie: Chocolate, white and choc/white roan

Saffron

Colour—deep lemon colour carried down to skin

Shape—short and cobby, deep broad shoulders

Coat—short and silky

> Ears—rose shaped and drooping
> Eyes—large, bold and *red*
> Condition—

Dark–eyed golden

> Colour—rich golden shade, ears and feet to match
> Shape—short and cobby, deep broad shoulders
> Coat—short and silky
> Ears—rose shaped and drooping
> Eyes—large, bold and as dark as possible
> Condition—

The crested varieties of these guide standard cavies may also be shown in guide standard classes, and are now, since the 1979 meetings of the R.V.C.C., C.C.C. and the B.C.C., catered for by the R.V.C.C., though the C.C.C. will put classes on for them at their stock shows.

THE UNSTANDARDISED CAVIES

These are the varieties which are waiting for a guide standard, but have been approved by the R.V.C.C.

Rex

This is the foremost one at present, and a guide standard has already been presented to the British Cavy Council at their meeting in October 1979. It was however decided to leave it for another year, until more progress is made. The Rex cavy is one of the most exciting of the new varieties of recent years and is now increasing in numbers, and improving in quality. It is already standardised in Canada, where they are called 'Teddies'. The Rex cavy's coat appears woolly. When they are babies it is curly. The whiskers also curl. The coat should be dense and short.

Rex bred to Rex will breed true, but it is also useful to breed with Rex Carriers.

One of the main faults to look out for is a "saddle" where the hair lies wrongly on the back giving a dip to the shape of the cavy.

Buffs

Many people consider these as dark creams, but buff bred to buff breed true. They are dark eyed. They can be confused with Saffrons, but the Saffron is red–eyed.

Sable

A shaded cavy, chocolate coloured on the back, and shading gradually to beige or cream on the belly. They can be dark, medium or light. They are very rare, but do appear at R.V.C.C. shows from time to time, and are very attractive.

New Varieties

New Varieties crop up from time to time, but very few of them ever make the grade. They can be shown only in New Variety classes at the R.V.C.C. Stock Shows. In the past year there have been no new varieties but new colours in the Argente range including saffron and lilac with silver ticking.

In Canada, they now have satin cavies, but they have not yet been seen in this country.

CHAPTER 16

The Show Scene

If you decide that you would like to show cavies, the first thing to do is to go along to your local show—have a good look round, and talk to the fanciers before deciding which variety you would like to take up. If you already have a cavy which, you think, is worth showing, take it along and obtain an opinion on it. Do not enter it in a show until one of the experienced fanciers has advised you. The fortnightly paper, *Fur and Feather**, which you can order from your Newsagent, will give you details of the shows. *You do not need to belong to any Club to show your cavies,* but probably as time passes, you will wish to join:—

1. Your local club
2. The Specialist Club for your breed
3. Your area club and/or the
   National Cavy Club

As a member of these clubs, you can compete for special trophies and diplomas etc. at local shows or at specialist club shows. In your local club, you will always be welcome to help at the shows in one way or another, which can be a very satisfying pastime, and a way of making new friends.

**Fur and Feather* is no longer produced, but there are magazines which deal with individual small animals; e.g. *Rabbits, Cavy Magazine.*

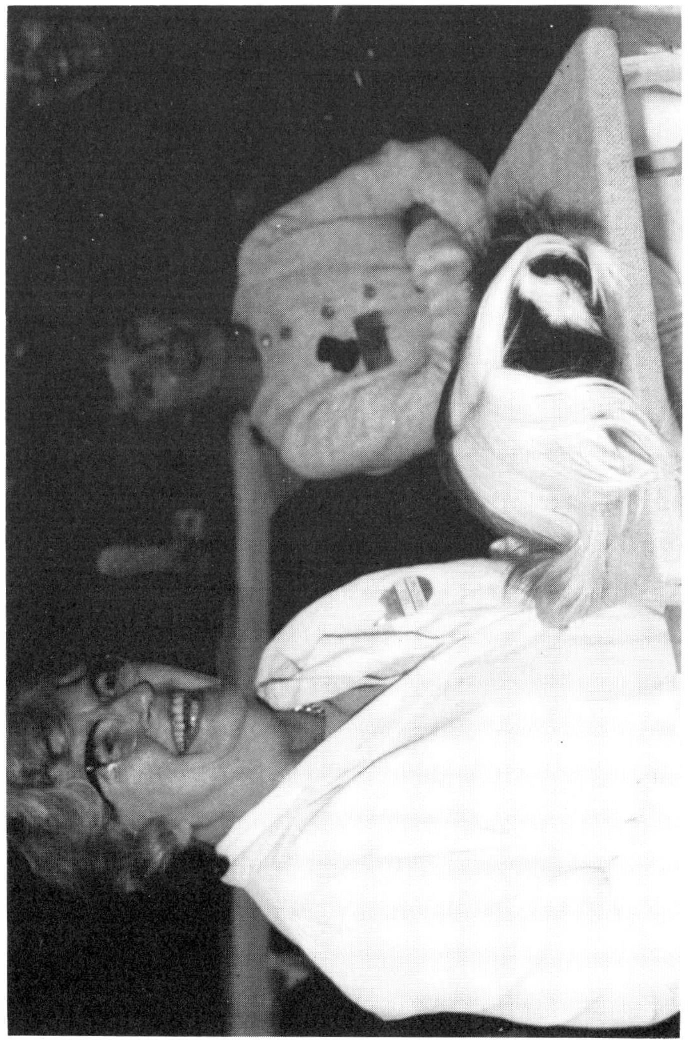

Figure 17 Mrs. Isabel Turner the Author enjoys a joke while judging at the London Championships Show.

THE CAVY CLUBS

The specialist clubs and the area and national clubs generally move their stock shows around the country to be within reasonable reach of as many members as possible. The secretary of the local club concerned does most of the preliminary work, but the specialist club secretary, or someone delegated by him, is usually there on the day, to do the secretarial work.

The area clubs cover an area of the country and cater for all breeds. The rules vary from club to club. The Southern Cavy Club covers the area south of a line through Birmingham, and does not take members from the north of that line; The Midland Cavy Club takes members from anywhere, usually fanciers who like to show in the Midlands. The Scottish Cavy Club also takes members from any part of the country, but it is based in Scotland. The National Cavy Club is the parent club and is nationwide. These Clubs, along with the Specialist Clubs, issue their members with a Year Book which contains useful and interesting information.

YOUR FIRST SHOW

Now you have some show standard cavies and you want to enter them in a show. Study your show advertisements in *Fur and Feather*. Some give a closing date for entries, which means that the entries (on paper, not the cavies!) must be in the hands of the secretary by that date. Some say, "Fees must accompany entries", but some secretaries accept entries by telephone, and you pay on arrival at the show—but, once you have given the secretary your entries, you must pay your entry fees, even if, by some misfortune, you are unable to take your cavies to the show.

It is usually the larger shows which require the entries before the event, because of the paper-work involved. Most of the smaller shows have "entries up to time of judging", which means that you take your cavies along to the show and enter them on your arrival there.

You may also read in your show advertisement "P.M. (i.e. prize money) 75%" This means that in each class, seventy-five per cent of the entry fees is paid out in prize money. You are usually very fortunate if your prize-money covers what it has cost you, but I hope you have taken cavies up as a hobby—*not* as a business. Some larger shows have guaranteed prize money which is paid regardless of the number of entries, but these are becoming fewer.

SHOW SCHEDULE

Now, we come to the show schedule, and the following is a typical schedule, followed by explanations:—

1. Self white or cream ad. *(adult)*
2. Self white or cream 5/8 *(five to eight months old)*
3. Self white or cream U/5 *(under five months old)*
4. Self black ad.
5. Self black 5/8
6. Self black U/5
7. Self A.O.C. A.A. *(any other colour: any age)*
D. 8. Self Challenge *(a duplicate class for any colour any age Selfs, which must have been entered in one of the previous classes)*

9. Abyssinian—A.C.Ad.
10.   ditto —A.C.5/8
11.   ditto —A.C.U/5
12. Agouti—A.C. Ad.
13.   ditto—A.C.5/8
14.   ditto—A.C.U/5
15. Tort and White A.A.
16. Dutch A.C. A.A.
17. Himalayan A.C. A.A.
18. Peruvian A.A.
19. A.O.V.A.A. *(any other variety, any age)*
20. Unstandardised A.A., not to be duplicated.

D.21. Non-self Challenge *(any variety bar Selfs)*
D.22. Sow A.V.A.A. *(any variety, any age)*
D.23. Boar A.V.A.A.
D.24. Ad Challenge *(open to all standard adults)*
D.25. 5/8 Challenge
D.26. U/5 Challenge
D.27. Grand Challenge *(open to all standard cavies in the show)*

THE ENTRY FORM

You now have to make your entry form out: for example, if you have an adult Self White sow of "High quality" which you wish to enter "Right Through" and a 5/8 Abyssinian boar which you are not too sure about, this will give you the idea of how they should be entered:—

1. Self White sow adult in Classes 1, D.8, D.22, D.24, and D.27.
2. Abyssinian boar 5/8 in Classes 10, and D.21.

At a 10p. per class, the cost of entering the self would be 50p. and the Abyssinian 20p.

A cavy must be entered in its breed class, before it can be entered in any of the duplicate classes. If there is no named breed class for your cavy, then it must be entered in the A.O.V. (any other variety) class. Many shows do not put the letter "D" before the number to signify "duplicate classes", but nevertheless, the rule still applies, and you soon learn which classes to enter. At shows where there are no 5/8 classes, any cavy over five months old must be entered as an adult, but it cannot be entered in 5/8 *and* adult classes at one show.

You will notice that the Selfs form a large section, and indeed there are usually about as many Selfs at a show as the total of other varieties, so very often there is one judge for Selfs and another judge for all the other varieties. They then combine to judge the classes which contain both Selfs and Non-selfs.

PRIZE STRUCTURE

When you are awarded a prize card, sometimes it is difficult for a newcomer to know what it means; 1st, 2nd and 3rd, are clear enough, and these are the cards with which you receive a prize. Then there is a "Reserve" card which means "4th"; V.H.C. (Very Highly Commended) which is 5th; H.C. (Highly Commended) being "6th"; and C.(Commended) being "7th". However, at many shows, cards up to "Reserve" only are given.

Some shows have a "Sportsman's Class". How this class is won may vary from show to show. The usual way is that the "Commended" or 7th, wins First Prize;

"Highly Commended" wins Second Prize and "Very Highly Commended" wins Third Prize.

TRAVELLING BOXES

To carry your cavies to the shows you need a "travelling box". Do not worry if you do not have one when you go to your first few shows; but have a look round and see the various designs of boxes which other fanciers are using. If you are good at carpentry, you can make your own, or there are a few fancier-carpenters who make them to sell. Not being a carpenter myself, I will not give instructions on making them, but I will give the basic requirements. A box may hold one, two, three or more cavies. Each cavy must have a separate compartment, large enough to allow him to sit in it comfortably, not less than 6in x 10in x 8in high. Usually, the partition between the compartments is removable to make cleaning easy, and if you want a larger compartment at any time, the partition can be removed. Peruvians and Shelties generally need a larger compartment.

The base of the box should be of wood at least half an inch thick—thinner would soon warp. Ventilation is important. Many boxes have circular holes—about two inches in diameter—at either end of the box, with a removable partition about half an inch away to screen the cavies from draught. In hot weather it is a good idea to remove the end partitions to give the cavies more air. The lids of the boxes also have ventilation, the most popular being a two level lid which admits air between the levels. Finally, you need a carrying strap, usually leather, which is affixed to each end of the box, and

Plate 4   *Top*   American Crested (Mrs Isabel Turner).
         Bottom   Silver Agouti (Mrs Pat Wood).

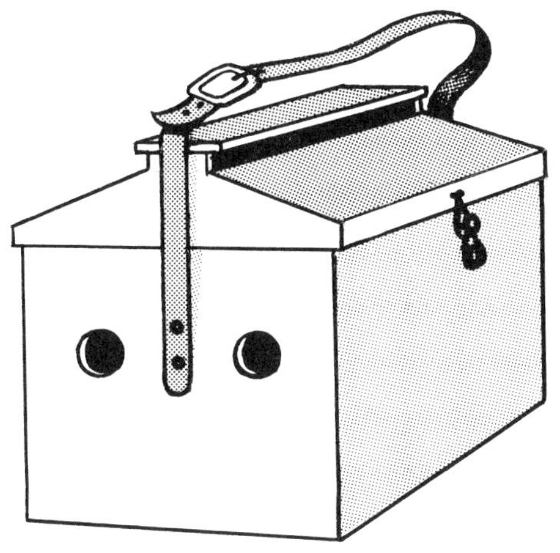

Show Carrying Box (shut)

End view of Box Lid
(when a flat topped box is made).

Side view of Box Lid
(when a flat topped box is made).

fastens on top by a buckle.

A word of warning—in cool weather and for short journeys, you can carry the boxes in the boot of your car, but in hot weather, or over long distances, you might reach the show with dead cavies in your boxes. In such circumstances, try if possible to have the boxes in the back of the car, and if possible, the lids open.

SHOW STANDS

Peruvians and Shelties require a stand at shows, and this is brought to the show by the fancier, not being supplied by the show authorities. The stand is like a little stool about six inches high and no more than sixteen inches square covered with hessian. Gone are the days when Peruvians are shown sitting on velvet cushions—yes, this did actually happen!! The Peruvian Cavy Club frowns on this practice and asks that Peruvians be shown on a hard flat surface. The necessity for these stands will be obvious when you see a Peruvian or Sheltie in full show coat. They are brushed out on their stands and carried to the judges' table still on their stand. If they were carried any other way, their coats would become hopelessly tangled. Because of the difficulty of keeping their coats just right, it is usual for the Peruvians and Shelties to be stewarded at the shows by their owners—or failing that, then they must be handled by a fancier who is familiar with the grooming routine. So, if you are asked to fetch a cavy from its pen, and find that it is one of these two varieties, do not touch it, but call a more experienced fancier who will find the owner.

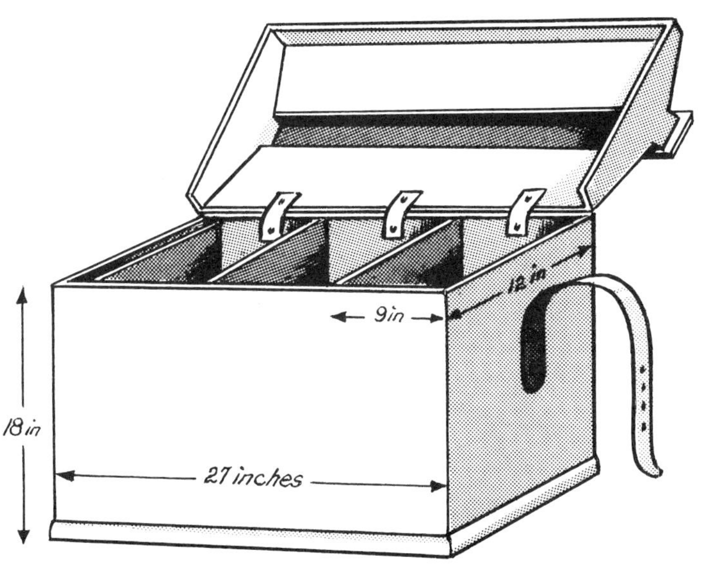

Show Carrying Box (open)
Three compartments

STEWARDING

If you are asked to steward at a show, it means fetching the cavies to the judges' table, and returning them to their pens afterwards. The cavies must be handled carefully, and it is usual to ascertain the sex of the one you are to fetch. When you reach the judges' table, give the cavy's number and sex to the book-steward; make sure the cavy has an ear-label with its number on, but if it does not have an ear-label, the book-steward will supply one. Carry only one cavy at a time, and do not smoke while stewarding.

Many newcomers are nervous of book-stewarding for the judge, but it only means acting as secretary. The book steward writes down the remarks which the judge asks him to, against the pen number of that particular cavy. He also enters the cavy's position in its breed class, against its pen number in the duplicate classes. Most judges are very patient with someone who is new to book-stewarding, and it is an excellent way to learn more about the cavies and how to judge.

BATHING YOUR CAVIES

About five days before a show it is usually necessary to bath your cavies. All cavies need a bath occasionally, but show stock need bathing more often. Use a baby shampoo and be sure to rinse the cavy well, using a jug of luke-warm water. Dab the surplus moisture off the cavy with a towel, and finish off either before the fire or with a hair-dryer. Some cavies love being bathed, and basking in front of the fire or under a hair-dryer, but others hate it. It is necessary to bath the cavies a few days before the show to give them time for the natural oils to return to the coat; therefore, do not leave it too late.

Peruvians and Shelties in show coat have to be bathed very carefully to make sure their feet do not catch in their coats. Do a little grooming while the cavy is wet—just enough to make sure there are no tangles, but be sure they are quite dry before you put their wrappers in again, and watch them carefully while they are drying, as they are quite capable of biting away some of their coats in their efforts to assist the drying process. The cavy's nails may also need trimming; use a nail clipper and be careful not to cut too near the quick or they will bleed. Above all, do remember that your show stock should be groomed regularly, and not just for a few days before the show.

CHAPTER 17

Improving Your Stock

MAKING A START

Before you think of buying stock, you should go round the shows, and, when you have decided which breed is your choice, look well at the winners, so that you have in your mind, a picture of a good specimen. You will find that some breeders will sell you good quality breeding stock a little over pet prices, while others, whose stock may or may not be as good will charge exorbitant prices—so price is no guide to the quality of the stock. Get in touch with the secretary of the specialist club for the breed of your choice, and he will give you the addresses of reliable breeders.

The boar you buy must be the best you can get, and if you can also get sows of equal quality, then you are off to a good start. The notes which follow deal with selection of the different varieties.

Selfs

Use a good solid boar, of good colour, high shoulders, large eyes with as much space between the eyes as possible. Look also for good undercolour. The sows are generally of better shape than boars and have a finer,

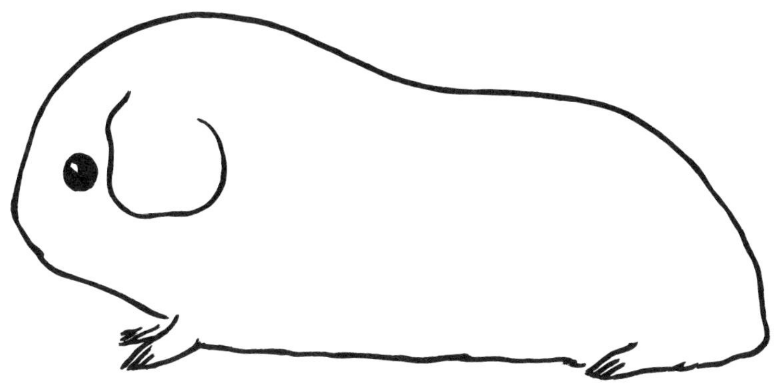

THE RIGHT SHAPE
Shape in a self. Note the high shoulders,
large eyes—well apart. Drooping ears.

THE WRONG SHAPE
No depth in shoulders. Small upstanding
ears. Small eye. Narrow, long head.

softer coat. Blacks, Whites and Creams have better type than the other colours, so discard any with long narrow heads, small eyes and pricked up ears. No matter how good the colour, **Black** youngsters with a sprinkling of red hairs generally lose those as they grow, and often have the best colour as adults. They also make good stock for breeding colour.

In **Whites** avoid those with a creamy or yellow tinge. For breeding stock choose those that are "whiter than white", combined with good shape.

With **Creams**, breeding the desired pale cream to pale cream continuously will produce youngsters which are *too* pale, so keep one or two buffs to breed with the creams. You will usually breed a few buffs in your cream litters so don't discard all of them.

The other colours of Selfs are generally not such a good type as the Blacks, Whites and Creams, although they are improving and some good ones appear. Breed with the best shaped animals you produce. If you have a gradual improvement over several generations, you are doing fine. In most colours you will be surprised at the different shades in one litter. When picking your best, don't forget that youngsters are darker than they will be when adult so make allowance for this fact.

In **Reds**, it should be noticed, that a little white—such as a white foot, for some obscure reason, are the best for breeding with—although a white foot would make a red too faulty to show.

Abyssinians

Breed with stock with good strong coats with good ridges and depth. If you are trying to breed self-coloured Abyssinians, you will find that they are generally soft in coat, but by mating Blue Roan to Black, Strawberry Roan or Brindle to Red, this helps to make the coat harsher. This advice was given to me many years ago by the late Harry Badcock, when I was trying to breed black Abyssinians, and it did work, but colour in Abyssinians is just a personal preference.

Agoutis

I have already mentioned the use of the black dilutes which appear in Silver litters. These are very useful for breeding with Silvers, if you find that the Silvers are getting too light.

With Golden Agoutis choose breeding stock from those that have really rich red-gold colour—not yellow or brassy.

Dutch

Here is a difficult one. You are just as likely to breed good ones from mismarked stock, as you are from good stock, but to my mind, what goes in must come out, be it several generations later, so keep careful breeding records—and for Dutch try doing little drawings of the markings of stock you intend to breed from and study them from generation to generation. Where you find a fault persisting then breed them to stock which excels on that particular point. Do not breed with poor coloured Dutch; e.g. a red Dutch should not look like washed out golden.

Tort and Whites

Much the same remarks apply as for Dutch but if your boar lacks one colour—for example, red on the left side, then try breeding him with a sow which has too much red on that side—and so on. Look carefully at the head-markings and avoid pairing two that both have Dutch heads; for example, if one has a red Dutch head pair it with one which has a solid black head. Look at the demarcation line down the belly. It should be as straight as possible. Avoid breeding with stock with brindled patches. But in spite of all this advice it is quite likely that someone will go against all this advice, and breed a "flyer"!!

Himalayans

Useful breeding stock are those dark-backed sows, which are too dark on the body colour for showing—provided your boar excels in body colour.

Do not use for breeding those that have flesh coloured patches on the pads of the feet, but remember Chocolate Himalayans *have* dark flesh coloured pads. Lighter patches are still visible on them, however. Some breeders use a Self-white cross to improve type in their Himalayans. My advice is "Don't". If you must outcross use a Self-black. My own feelings are that the Himalayan is a beautiful cavy which is a different type completely from the Selfs. So why try to make it like a smudgy faced Self-white?

Peruvians

Choose your breeding stock, which has a good dense coat, especially on the shoulders—and a good frontal. Avoid those with an excess of body rosettes. Look for those with flat broad faces, and not those with long narrow faces.

Shelties

Similar remarks as for the Peruvians but of course

thay have no frontal, and no rosettes. Again go for thick dense coats and broad faces, but also look for a coat that grows in length—at the rate of about an inch per month.

Cresteds

You will find that you have to breed occasionally with Selfs to improve type and colour. Although blacks, whites and creams are now established with good type. I have even heard of a breeder who is using Cresteds to improve the type in their Selfs. You will get roughly 50% Crested whether you breed Crested to Crested or Crested to Self.

You may find that Crested to Crested sometimes produce faults such as double-centred crests, a ridge down the back or even a crest in the middle of the back, and occasionally side whiskers—which are little tufts of hair sticking straight out, under the ears. It is best to avoid breeding with those faulty ones. They may not pass the fault on, but there is always the chance that they might. It is unfortunate that these faulty ones are usually beautiful cavies apart from the fault.

When breeding with American Cresteds you may find red or golden hairs creeping in—no matter what colour you are breeding—with the exception of Creams. If you are trying to produce new colours, you may have no alternative but to use these "brindleys" in your first generation, but you will find that they clear in a couple of generations.

In any variety, don't overshow youngsters if you want to breed with them or you may have difficulty in getting them to breed. If a youngster is shown two or three times before breeding it is enough.

CHAPTER 18

The Show Secretary

Sometimes a person has to suddenly take on the task of show secretary, and has no previous experience. It is hoped that the secretary has a good committee to help, but in most cases, the committee look to the secretary for instructions.

THE ROUTINE WORK

Taking over from a previous secretary means taking over "the books" and it can look quite formidable. The majority of secretaries also act as treasurer, so a visit to the bank is necessary to let them know of the change, and to give them the usual sample of your signature. It will be necessary, of course, to have the former secretary—or the chairman or president of the club to vouch for you. You will have to keep accounts of money coming in and going out, so that the books may be audited annually—so of course you will need an accounts book—and a rough one from which you can copy the items into the auditor's book. Also keep a postage book, and it is wise to enter every letter posted—even if you have had a stamp for a reply. It then serves as a means of checking whether or not a certain letter was posted. Just put P.P. (post paid) opposite those in the book, who have sent a stamp. You will of course have a members address book, with columns to tick off when subscriptions are paid, and a receipt book. If your club owns any cups or trophies, you need a book in which to record, who is holding them—and a "minutes book" to record your meetings.

Those are the minimum requirements—of course there are others, as will be obvious.

ORGANISING SHOWS

The main part of the secretary's work is organising shows—whether they be the large open shows—or your local show which may be as often as once a month. The committee should meet to make all the arrangements, but there is a lot for the secretary to do at home.

One of the first tasks is to engage a judge—or judges. This must be done many months before a show, as most of the top class judges are booked up very quickly. It is wise for your committee to have a second or even a third choice—just in case your first choice cannot accept.

For a show of any importance, your choice of judge should be one who is on the panel of one of the main cavy clubs. For the smaller shows it is permissable to ask someone who has a fair amount of experience in the cavy fancy. It is definitely *not* a good idea to ask the rabbit judge to judge the cavies—unless of course he is also a cavy fancier. Your choice of judge *is* important, because good, fair judges attract entries to your show.

Write to your judge(s), giving date of the show, time of judging—and if there is a second judge tell him who it is. Tell him also whether you would like him to judge Selfs or Non-selfs and ask him what his terms are. Enclose a stamped addressed envelope for a reply. Most cavy judges charge only a very small fee—or none at all, but your club must be prepared to pay all his travelling expenses, and overnight accommodation if judging starts early, and he has a long way to come. If it is an all-day show, you must make arrangements to supply him with a meal, and cups of tea or coffee at intervals. The refreshment ladies of your club are worth their weight in gold to everyone who is working at the show.

ENTERTAINING A SPECIALIST CLUB

If you wish to entertain any of the specialist clubs, watch in the specialist clubs notes in *Fur and Feather*. You will see, then, when the specialist secretaries ask for offers to stage their shows, so you can then write, giving them your terms—for example, you supply penning, meals for the judge and specialist club secretary, and they pay their own judge and supply their own prize cards, and pay for their share in your advertisement in *Fur and Feather*. Sometimes you may have to impose a penning charge, or a share of the rental of the hall; etc. but remember, it is to your advantage to entertain a specialist club. Their members will come to the show from many parts of the country, and the majority of them will enter your show as well. This should all be settled many months before the show. Do not, however, take on too much. Don't think of entertaining one of the large specialist clubs if you only have a tiny hall, and if you apply for the Peruvian—or Sheltie Club, remember that they also require quite a deal of table space for grooming on. You know what space you have—the specialist secretary does not, always, and he could accept your offer in good faith—and regret it afterwards.

SPECIALIST CLUB SUPPORT

You should also apply to the specialist clubs for their support in your show—which means that they may award Diplomas, Certificates of Merit, Rosettes or other specials for their members to compete for at your show. A few of the specialist clubs give automatic support which means that you need not apply for it. In this case, the exhibitor, who wins with that particular breed, sends his first prize card to his secretary, who will then send him his "special". However, in the majority of cases you have to apply for support. Write to each club secretary giving him particulars—the date of your show, the name of the judge and what classes you intend putting on for this breed. Enclose a stamped addressed envelope for a reply and after the show send him a list of the winners of his specials.

The same procedure is required if you want support from, or to entertain any of, the area clubs such as the Southern Cavy Club or the Midland Cavy Club—or the parent club—the National Cavy Club.

THE SCHEDULE

Now we come to the Show Schedule. This largely depends on which varieties are popular in your area, or on which specialist club you are entertaining. Consider these aspects when deciding what classes to put on for each breed. Your committee should meet to arrange the schedule, the entry fees, prize money; etc, but it is up to you, as secretary, to have either a copy of a previous show schedule—or a list of probable classes.

Start with your Self classes, followed by your Self duplicate classes. Then go on to the Non-self classes. Be sure you cater for every standardised breed—if there are no classes for some of the breeds, then you must put on an A.O.V. (any other variety) class. The standardised Rare Varieties may be entered in this class, if there is no Rare Varieties class. If you wish, you may put on an Unstandardised class, but cavies entered in this class should not be entered in Duplicate classes.

SHOW ADMINISTRATION

Don't forget to check that you have sufficient prize cards for your show, and write to cup-holders asking them to return the cups in time for the show. As the time for the show draws near, you must get your show advertisement in *Fur and Feather*. If it is a small show with entries up till time of judging, then it is sufficient to put in a small advertisement stating place, date, time of judging, name of judge—never "judge applied for" without a name—this could lose you valuable entries.

If it is a big show, you will require to have your entries in a few days before the show, or you will have more clerical work than you can cope with on show day. In this case it is wise to put the schedule in *Fur and Feather*, so that fanciers can enter from it. State name of show, place, date, entry fees, prize money, name of judge(s) and time of judging. Then the schedule with each class

DIAGRAM No. 1

CLASS 1. SELF WHITE AD.		WINNERS	
Pen No.		Duplicates		
1	John Smith, 2 Green Cres., Southwich	21, 22		
2	John Smith " " "	20, 22		
3	A.N. Other, 100 Smith St. Northwich	20		
4	A.N. Other " " "	—		

A sample of a page from an entry book.

numbered and the closing date for entries. It is wise, too, to say entry fee with entries. A few of the agricultural shows and the City shows do send out printed schedules by post.

Send your advertisement to *Fur and Feather* about a month before the show. About this time, too, write to your judges, enclosing passes to get into the show ground (if these are necessary) and a car pass. These you will have received from the Agricultural Society or whoever is running the show. If it is a show in a hall being run by your society then these should not be necessary, but write to your judge anyway, saying you look forward to seeing him at the show.

Before your entries start coming in prepare your entry book—A4 size is best. A loose leaf book is ideal, in case you need extra paper. Leave a narrow column on the left edge for the Pen number. The three quarters of the line for the exhibitor's name and address, and two columns at the right-hand edge—one for the duplicate classes in which that cavy has been entered, and the last column left blank meantime—is for the winners (see diagram No. 1). At least one page is allocated for each class. On the pages for the duplicate classes, the column for duplicates should be changed to one for the number of the breed class. If you are entertaining a specialist club, you must make out a separate entry book for their section and give it to their secretary when he arrives at the show. As the entries arrive fill them in the book, but omit putting in the pen number at this stage.

If it is a show where you take entries till time of judging, it is usually sufficient to have a sheet of paper

DIAGRAM No. 2

Pen No.	NAME	CLASSES									
		1	2	3	4	5	6	7	8	9	10
1	John Smith	✓								✓	✓
2	John Smith	✓								✓	
3	A. N. Other		✓								✓
4	A. N. Other		✓							✓	✓
5	Mrs. Brown				✓						✓

A sample of how to take entries when entries are allowed up to time of judging.

DIAGRAM No. 3

75% PRIZE MONEY   **10p ENTRY FEE**

NUMBER OF ENTRIES IN CLASS		1	2	3	4	5	6	7	8	9	10	11	12	13	14	15	16	17	18	19	20
Prizes in pence	1st	8	10	12	15	18	21	24	27	30	33	36	39	42	45	48	51	54	57	60	63
	2nd		5	7	9	12	14	16	19	22	24	26	29	32	34	36	39	42	44	46	49
	3rd			3	6	8	10	12	14	16	18	20	22	24	26	28	30	32	34	36	38

15p ENTRY FEE

NUMBER OF ENTRIES IN CLASS		1	2	3	4	5	6	7	8	9	10	11	12	13	14	15	16	17	18	19	20
Prizes in pence	1st	12	15	18	24	30	36	42	48	54	60	66	72	78	84	90	96	102	108	114	120
	2nd		7	9	13	16	20	23	27	30	33	36	40	43	47	50	54	57	61	64	68
	3rd			7	8	10	12	14	16	18	20	22	24	26	28	30	32	34	36	38	40

GUIDE TO 75% OF ENTRY FEE PRIZE MONEY

If you are paying out 75% of the entry fees in prize money you may find the above charts useful. There is no rule about how the money should be divided and some may wish to give a little less for 1st and more for 2nd, etc., but these charts could be used as a guide.

with columns for the classes, with a column for pen number and another for the name of the exhibitor. Tick the column of each class entered. You need to put the pen numbers on at the start, so that you can give each exhibitor his pen number as he enters. This will mean that the varieties will not be in sequence, but it is the best way for this type of show (see diagram No.2).

In the past, all exhibitors entering a show, for which the entries closed a few days before the show, were sent a label to attach to their boxes, with instructions to write their name and address' and nearest station (if the cavies were sent by rail) under the flap. This flap was addressed to the show. On arrival the flap was torn off and the box was ready for the return journey. These labels have now been discontinued except for some of the very large shows, such as the Bradford Championship and the London Championship (even these may be discontinued).

If your show is at an Agricultural show or a City show you will have to send some of your exhibitors entry passes and perhaps car passes. These are usually restricted to those exhibitors who have spent more on entry fees than they would to get into the grounds, and car passes may be for *unloading* only.

Be sure to send passes to the secretary of your guest Specialist Club and their judge. You should, of course, make your entry closing date early enough to give you time to send out your exhibitors' passes and for them to get to their destination.

Once your entry closing date has passed, you can fill in the pen numbers in your entry book. By leaving it till this time, you get all the same varieties of cavies grouped together.

If you find that your exhibits outnumber the number of pens available, you must make arrangements to hire others. Perhaps some of your neighbouring clubs will oblige. You must make transport arrangements for taking all pens, staging; etc., to the show and returning them later. Also arrange supplies of hay and sawdust and, if necessary, food. The time for the show is drawing very near, and you will be very busy on show day, so anything you can do now to minimise the work on show day, will help.

First of all you can prepare envelopes—each one with the name of an exhibitor. Inside have a slip of paper with the pen numbers of each of his animals, and his numbered sticky ear-labels, which he sticks on his cavy's ear. If he has several animals entered be sure to make it clear exactly which pen number belongs to which animal. If you have the envelopes in a box in alphabetical order it is easy to pick each one out as the exhibitor checks in at the show.

You can also prepare envelopes with the prize money, 1st, 2nd and 3rd for each class. If the prize money is 75% you will have to claculate each class (see diagram No. 3). Then you can make up your prize cards with the information you have available—date, name of judge, number of exhibits in each class. Then all you will have to add at the show is the exhibitor's name and pen number. You will, of course, have the judges' book made up ready to hand to him to start judging (see diagram No. 4).

The day before the show, your helpers will be busy

DIAGRAM No. 4

CLASS 1. SELF WHITE AD. No. of Cards *4*		CLASS 1.	Secretary CLASS 1.	
Pen Nos.	Judge's Comments	Award	No.	Award
1		*3*	*1*	*3*
2		*4*	*2*	*4*
3			*3*	
4			*4*	
5		*1*	*5*	*1*
6			*6*	
7			*7*	
8			*8*	
9			*9*	
10		*2*	*10*	*2*
		Signature	*Signature*	

A sample of a judge's book. The dotted line is perforated and is torn off and returned to the secretary as each class is judged. The judge retains the book and from his comments can write his report for *Fur and Feather.*

putting up pens, judging tables, etc. Take your entry book with you and supervise the numbering of the pens. Ensure that Peruvians and Shelties—especially adult cavies have large pens with large doors, and as far as is practical try to put youngsters of other varieties in pens which they cannot squeeze out and escape from. Some of the babies may be quite small—and quite venturesome.

THE BIG DAY

When the day of the Show arrives, be there early with all your necessary books, papers, prize cards, prize money, a good stock of pens and pencils, rosettes, cups and other equipment. Be seated at your table ready to check in all the cavies as they arrive. Tick them off in your entry book so that you will know if any are absent and hand each exhibitor his previously prepared envelope containing ear labels and other instructions. It is usual to wait a short time for latecomers, but don't give them too long or you may find they don't try to be on time at your next show. Sometimes the judge will judge other classes first, and leave the classes with absentees till later, but if the entry is large there is a great deal of work and, if possible, judging should start on time.

Make sure your judge has plenty of stewards to fetch and carry stock for him, and ask him if he would like a book steward—though some judges like to do their own writing. When he has judged each class he will send a steward to you with the tear-off strip from the judging book, and from this you will be able to write up the prize cards. Clip these strips to each page of the entry book, and keep carefully for a few weeks in case you get any queries. It is best to keep them until after the judge's report appears in *Fur and Feather*.

Be sure the judge has a list of specials so that he can nominate the winners. The winner of the Grand Challenge is *not* automatically Best-in-Show. The judge may nominate as Best-in-Show, a cavy which has not been entered in the Grand Challenge.

You can also, as you get the results from the judge, fill in the names of the winners in the *Fur and Feather* Show report sheet which will have been sent to you. At the top fill in the particulars—name of show, date, judge's name, and number of entries. Then before he goes home give this report sheet to the judge. He will write his report on each of the winners and send it to *Fur and Feather*.

And don't forget to give him his fee and expenses.

Before the exhibitors go home with their cavies, they will bring their prize cards to you and you will give them their envelopes with the prize money they have won, and any trophies they have been awarded.

After the show, there is a lot of clearing-up, and unless you all have constitutions of horses, you and your team will feel shattered, but you should not have very much more writing to do. You may have some prize money to send out. There is always someone who goes without it, and you will have to notify the specialist clubs—who have won their specials—and you will have to keep your accounts straight.

Then it is time to start thinking about your next show!

CHAPTER 19

The Specialist Club Secretary

Very few fanciers realise what a vast amount of work has to be done by a Specialist Club secretary or the secretaries of the large Area Clubs and the National Cavy Club. It is a totally different job to the job of the show secretaries except, of course, that like the show secretary's job it is unpaid. They all do it for the love of the Fancy.

The Specialist Club secretary is very much on his own. His committee is scattered around the country maybe hundreds of miles apart, so committee work must be done by letter. The members, too, are scattered—not only all over the country—but all over the world.

A LABOUR OF LOVE

The secretary must be prepared to spend hours every week just replying to letters. Correspondence does appear to be the most time consuming aspect of the job. There are letters from members—and non-members— asking for advice on all sorts of problems. Some would need a book to answer them—so they need very long replies. Occasionally there are letters of complaint such as "I bought some stock from one of your members, and it is not up to standard. What are you going to do about it?". Well, there is nothing a secretary can do. If the writer had written before buying stock, the secretary would probably have been able to recommend a likely source of good stock, but, as membership is so scattered, the secretary cannot know all his members—or the quality of their stock—and no secretary will recommend anyone unless he knows that the vendor has good stock and will treat the buyer fairly. Sometimes, problems in an area can be passed on to a committee member who lives in that area—so the scattered committee does have its good points.

There are letters from children with pets, letters from people asking how to go about exporting stock, and letters from abroad asking about importing stock to their country. Letters from members claiming diplomas and championships, from show secretaries asking for show support—some silly letters like "Where can I buy a cavy good enough to win a championship show, but I don't want to pay more than £1 for it". Yes, this actually did happen!! Secretaries don't know all the answers, but they do try to make a good job of answering all these letters.

THE STOCK SHOWS

The Specialist Club's Stock Shows may be in any part of the country—there are usually two or three per year. The secretary must try to attend. If he can't then he must delegate someone to represent him, and he must see that all necessary prize cards; etc, get to the show. He also has to arrange for the return of cups and trophies, which are to be competed for at the show.

THE ANNUAL GENERAL MEETING

The Annual General Meeting of the club must be arranged at some place where as many members as possible will be able to attend. This is usually either at one of the Stock Shows or at one of the big Championship Shows. If the Stock Show is at an Agricultural or City Show it is almost impossible to find anywhere satisfactory to hold the meeting. At a one-day show there is so much to do, and people who have come a long way are anxious to be on their way when the show is over, that it often means that the meeting is rushed and not very satisfactory. The best place is often at a two-day Championship Show. Because of the nature of the show, there is usually a large attendance so there will be a fair number of Specialist Club members. On the second day of the show, the judging is over, the atmosphere more relaxed, and that is when and where many Specialist Clubs hold their A.G.M.'s.

The agenda for the A.G.M. has to be prepared beforehand. It is usually at this meeting that the applications to entertain the Stock Shows are read out and considered, as well as all the problems that have cropped up since the last meeting, and any proposals from members. The accounts are also presented to the members, so they have to be audited just before the A.G.M.

ELECTION OF OFFICIALS

The ballot takes place in some clubs annually, but

owing to the high costs, most clubs now have a ballot every two years. Before a ballot can take place, the secretary must ask for nominations for officials, committee and judges. The nominees must be willing to stand. All the officials, committee and judges in office at the time are automatically on the list unless they indicate that they wish to retire. The secretary eventually lists all the names under the different offices, has the list duplicated, so that a copy can be sent to each adult member. The members are asked to signify their votes by crosses, and send it to the official scrutineer, who is not a member of the club. The judges for the stock shows are usually chosen by means of the ballot.

CLUB NEWS

The Club Notes in the Fancy magazines are the best way of keeping members and club officials in touch. The Year Book is another important channel of communication—often the secretary has to collect material, and put it together, get it printed and sent out to members. The Year Book contains lots of useful information plus lists of members, lists of cup and diploma winners, interesting articles, and advertisements. These Year Books are usually very much sought after.

There are many other duties of the Specialist Club Secretary—keeping records of all sorts of things, ordering Diploma cards, trophies, rosettes; etc., and trying to keep his members happy—he can't possibly please everyone all the time—but he does his best.

APPENDIX I

CLUB SECRETARIES

National Cavy Club: Mr P. Parkinson, 23 Union St. Slaithwaite, Huddersfield, Yorkshire HD7 5ED.

Southern Cavy Club: Mr C. Clouter, 56 Allison Rd. Bristol BS4 4PN.

Midland Cavy Club: Mr B. E. Berry, 62 Queens Rd. East, Beeston, Notts.

Scottish Cavy Club: Mr D. Rennie, 1 Westgate, Bucksburn, Aberdeeen, Scoland.

SPECIALIST CLUB SECRETARIES

English Self Cavy Club: Mr S. Heard, 24 Somerhill Rd., Welling, Kent.

Abyssinian Cavy Club: As above.

Agouti Cavy Club: Mr G. Fox, 50 Carr Manor View, Leeds LS17 5QA.

Peruvian Cavy Club: Mrs I. Turner, 1 Archdale St. Syston, Leicestershire.

Sheltie Cavy Club: Mrs R. Colley, 33 Jan Pallach Ave. Nantwich, Cheshire.

Dutch Cavy Club: Mr P. J. Dolphin, Windyridge, 9 Perryfield Rd., Bromsgrove, Worcs.

Tort and White Cavy Club: Mr R. W. Joynson, 75 Leysholme Cres. Wortley, Leeds LS12 4HH.

Himalayan Cavy Club: Mr A. Jackson, The Woodlands, Renwick Lane, Cheadle, Stoke-on-Trent, Staffs ST10 1TX.

Crested Cavy Club: Mrs J. Pickering, Fernhill Cottage, Rouncil Lane, Kenilworth, Warks CV8 1NN.

Rare Varieties Cavy Club: Mr C. Oxtoby, 111 Queens Road., Vicars Cross, Chester.

OVERSEAS CLUB SECRETARIES

American Cavy Breeders Association: Vern Emery, PO Box 416 Herrin, Illinois 62948.

Ontario Vacy Club: Mrs R. Eisel, c/o General Devliery, Main Post Office, St Catherines, Ontario.

Texas Cavy Breeders Association: Frank Simmons, 5006 Casa Oro, San Antonio, Texas 78239.

Western Australia Cavy Club: Miss S. Sumner, 23 Langham Gds. Wilson, W.A. 6107.

National Cavy Club of New Zealand: Mr A. Benson, 126 Waterloo Rd., Lower Hutt, N.Z.

Victorian Cavy Club: Mrs C. Black, 2 Parer St. Oakleigh, Victoria 3166, Australia.

APPENDIX II

FURTHER READING

Cavy Magazine
from: A. Waspe, 288 Baddow Road, Gt Baddow, Chelmsford CM2 9QX

Greenfood For Rabbits And Cavies – Practical Inbreeding
by W. Watmough
Cavies
by A. Sole
Available from the Book Department

Colour Genetics Of The Cavy – an Introduction
by Catherine Whiteway
Available from Mrs Whiteway, Langley Cottage, Lower Landley,
Wivelscombe, Somerset

The Rare Varieties Cavy Club have a library of books and papers,
particulars may be obtained from Mrs Whiteway.

Your Guinea-Pig
published by Colourmaster International, obtainable from Pet Shops.

"Pigs Isn't Pigs"
by Wynne Eecen, published by Favoretta Publications, Sydney, Australia.
Available from Box 7005, GPO Sydney, Australia.

Guinea Pigs
by P. Hutchinson
Available from the Book Department